LEARNING ICT with MATHS

Other titles in the Teaching ICT through the Primary Curriculum series:

Learning ICT in the Arts
Andrew Hamill
1-84312-313-4

Learning ICT with English
Richard Bennett
1-84312-309-6

Learning ICT in the Humanities
Tony Pickford
1-84312-312-6

Learning ICT with Science
Andrew Hamill
1-84312-311-8

Progression in Primary ICT
Richard Bennett, Andrew Hamill and Tony Pickford
1-84312-308-8

LEARNING ICT with MATHS

Richard Bennett

Routledge
Taylor & Francis Group

LONDON AND NEW YORK

Supplementary Resources Disclaimer

Additional resources were previously made available for this title on CD. However, as CD has become a less accessible format, all resources have been moved to a more convenient online download option.

You can find these resources available here: www.routledge.com/9781138414563

Please note: Where this title mentions the associated disc, please use the downloadable resources instead.

First published 2006 by David Fulton Publishers

Published 2014 by Routledge
2 Park Square, Milton Park, Abingdon, Oxon OX14 4RN
711 Third Avenue, New York, NY 10017

Routledge is an imprint of the Taylor & Francis Group, an informa business

British Library Cataloguing in Publication Data
A catalogue record for this book is available from the British Library.

ISBN: 1-84312-310-X
EAN: 978 1 84512 310 1

Typeset by Servis Filmsetting Ltd, Manchester

Contents

CD Contents   vi

Acknowledgements   viii

Introduction   1

1   Using programmable toys   9

2   Counting   17

3   Shopping   29

4   Exploring with directions   39

5   Symmetry and tessellation   51

6   Statistical investigations 1   69

7   LOGO challenges   83

8   Modelling investigations   95

9   Patterns and spreadsheets   111

10   Statistical investigations 2   123

Index   135

CD Contents

Resources and links (MS Word)
Certificate Read-Me (Text file)

Project 1
- Project 1 Completion Certificate (MS Word)
- Project 1 Completion 'Star' Certificate (MS Word)
- Project 1 Evaluation (MS Word)

Project 2
- Project 2 Completion Certificate (MS Word)
- Project 2 Completion 'Star' Certificate (MS Word)
- Project 2 Evaluation (MS Word)

Project 3
- Project 3 Completion Certificate (MS Word)
- Project 3 Completion 'Star' Certificate (MS Word)
- Going Shopping (Installation file)
- Going Shopping prices (MS Word)
- Project 3 Evaluation (MS Word)

Project 4
- Project 4 Completion Certificate (MS Word)
- Project 4 Completion 'Star' Certificate (MS Word)
- Terry the Turtle Evaluation Version by Kudlian Software (Installation file)
- Terry the Turtle file – turtle-islands.tt
- Terry the Turtle file – turtle-town.tt
- Crash (Installation file)
- Project 4 Evaluation (MS Word)

Project 5
- Project 5 Completion Certificate (MS Word)
- Project 5 Completion 'Star' Certificate (MS Word)
- Polygons Around a Point record sheet (MS Word)
- Rotational Symmetry e-worksheet (MS Word)
- Tessellation examples (Bitmaps)
- Project 5 Evaluation (MS Word)

Project 6

- Project 6 Completion Certificate (MS Word)
- Project 6 Completion 'Star' Certificate (MS Word)
- NNS information sheet (pdf file)
- Textease examples (data sheet and example of findings)
- Data files – colours – football colours (csv and tsv format)
- Granada Database files – colours – football colours
- Project 6 Evaluation (MS Word)

Project 7

- Project 7 Completion Certificate (MS Word)
- Project 7 Completion 'Star' Certificate (MS Word)
- LOGO challenges sheet (Word file)
- Solutions to challenges (HTML document)
- Solutions to challenges (SuperLOGO file)
- Solutions to challenges (Text file)
- LOGO procedures for the SCENE (Text files)
- LOGO files for the SCENE (SuperLOGO files)
- LOGO program for SCENE (Installation file)
- Project 7 Evaluation (MS Word)

Project 8

- Project 8 Completion Certificate (MS Word)
- Project 8 Completion 'Star' Certificate (MS Word)
- Example target spreadsheet (MS Excel)
- Example target practice (MS Excel)
- Example some practice (MS Excel)
- Recording sheet for Duck game (MS Word)
- Project 8 Evaluation (MS Word)

Project 9

- Project 9 Completion Certificate (MS Word)
- Project 9 Completion 'Star' Certificate (MS Word)
- Function machine (MS Excel)
- Number cruncher (MS Excel)
- More ideas (MS Excel)
- Match patterns sheet (MS Word)
- Pattern investigation worksheet (MS Word)
- Function investigation worksheet (MS Word)
- Project 9 Evaluation (MS Word)

Project 10

- Project 10 Completion Certificate (MS Word)
- Project 10 Completion 'Star' Certificate (MS Word)
- Year 3 stats datafile (MS Excel, tsv file, csv file, text file, Textease Database file)
- Class database (Textease Database file)
- Project 10 Evaluation (MS Word)

Acknowledgements

The author would like to thank the following for granting permission to reproduce the screen images and software files used in this publication:

2Simple for screenshots of *2Go* and *2Count*

Digital Workshop for use of the *Illuminatus* installation files on the CD

Granada Learning for screenshots of *Granada Database*

Kudlian Soft for screenshots of *Terry the Turtle* and the demonstration version of *Terry the Turtle* on the CD

Logotron for screenshots and use of the *SuperLOGO* installation files on the CD

Softease for screen shots of *Textease, Textease Database, Textease Turtle*

SPASoft for the screenshot from *Let's Go Shopping*

The author would like to thank Oxford University Press for permission to include the URL link to its online activities.

Introduction

This book is based on the belief that the integration of information and communication technology (ICT) and subject teaching is of benefit to children's development through the Foundation Stage, Key Stage 1 and Key Stage 2. It focuses on ICT in the context of mathematics and numeracy. By incorporating some of the powerful ICT tools described in this book in your planning for mathematics, the quality of your teaching and children's learning will improve. Similarly, by contextualising the children's ICT experience in meaningful mathematics projects, children's ICT capability will be enhanced and extended. *Learning ICT with Maths* is one of a series of ICT books: Teaching ICT through the Primary Curriculum. The core book for the series, *Progression in Primary ICT*, provides a more detailed discussion of the philosophy behind the approach and offers an overview and a planning matrix for all the projects described in the series.

The activities that are presented here offer practical guidance and suggestions for both teachers and trainees. For experienced teachers and practitioners there are ideas for ways that ICT can be developed through the areas of learning and the primary mathematics curriculum using ICT tools with which you are familiar. For less confident or less experienced users of ICT there are recommendations for resources and step-by-step guides aimed at developing your confidence and competence with ICT as you prepare the activities for your children.

The activities are related to the Foundation Stage areas of learning, the National Curriculum Programmes of Study (PoS) and the national framework for numeracy. In some activities, such as Project 2: *Counting*, Projects 6 and 10: *Statistical investigations*, the emphasis is on finding things out with ICT. Other projects, e.g. Project 1: *Using programmable toys*, Project 4: *Exploring with directions* and Project 7: *LOGO challenges*, offer opportunities for developing ideas and making things happen using ICT tools. Children are provided with purposeful opportunities to exchange and share information in Project 5: *Symmetry and tessellation* and Project 9: *Patterns and spreadsheets*. Throughout all the projects, ways in which children can reflect on their use of ICT or explore its use in society are identified. Although the projects are clearly located in mathematical contexts, several activities could be adapted for other subjects such as science or art.

The projects do not provide an exhaustive or definitive list of ICT opportunities in mathematics teaching. Instead, they are tried and tested sequences of activities, adaptable across the age-range, which ensure that high quality learning in ICT is accompanied by high quality learning in numeracy-related contexts. The projects are closely linked to relevant units in the Qualifications and Curriculum Authority (QCA) schemes of work for ICT and guidance is provided on how they can be used to supplement, augment, extend or replace units. Although the projects are not future-proof, they have been designed to take advantage of some of the latest technologies now available in primary schools, such as interactive whiteboards, internet-linked computers and digital cameras.

A note on resources

Investment has improved the level of resources for the teaching of ICT in primary schools in recent years. The arrangement and availability of resources, however, still varies greatly from school to school. Some schools have invested heavily in centralised resources, setting up networked computer rooms or ICT suites. Others have gone down the route of networking the whole school, using wired or wireless technologies, with desktop or laptop computers being available in every classroom. Some schools have combined the two approaches, so that children have access to a networked suite and classroom computers. This book does not attempt to prescribe or promote a particular type of arrangement of computer hardware, but does make some assumptions in relation to the management of those resources. These assumptions are:

⊙ The teacher has access to a large computer display for software demonstration and the sharing of children's work – this could be in the form of an interactive whiteboard (IWB), a data projector and large screen or a large computer monitor.

⊙ Pupils (in groups or as individuals) have access to computers for hands-on activities – this may be in an ICT suite or by using a smaller number of classroom computers, perhaps on a rota basis.

⊙ The school has internet access, and at least one networked computer is linked to a large display, as described above.

⊙ Pupils have access to internet-linked computers and the school has a policy for safe use of the internet.

⊙ Teachers and pupils have access to a range of software packages, including a web browser, 'office' software (such as a word processor) and some 'educational' software. Although this book makes some recommendations with regard to appropriate software, it also suggests alternatives that could be used if a specific package is not available.

The projects

Each project is presented using the following format:

⊙ a Fact Card which gives a brief overview of the project content and how it links to curriculum requirements and documentation;

⊙ detailed guidance on how to teach a sequence of ICT activities in a subject context;

⊙ information on pupils' prior learning required by the project;

⊙ guidance for the teacher on the skills, knowledge and understandings required to teach the project, including step-by-step guidance on specific tasks, skills and tools;

⊙ clear and specific information about what the children will learn in ICT and the subject;

⊙ guidance on how to adapt the project for older or more experienced pupils;

⊙ guidance on how to adapt the project for younger or less experienced pupils;

⊙ a summary of reasons to teach the project, including reference to relevant curriculum documentation and research.

National Curriculum coverage

The ICT activities described in this book are those which are most relevant to mathematics learning and hence not all areas of the ICT curriculum have been covered. The core text for the series, *Progression in Primary ICT*, shows how coverage of the ICT curriculum can be achieved by selecting the most appropriate subject-related activities for your teaching situation and how progression in ICT capability can be accomplished through meaningful contexts. Figure 1 provides an indication of the aspects of ICT which are addressed by the projects in this book.

Focus age groups for each project

Figure 2 provides an indication of the age group for which each project has been written. However, most activities can be adapted for older or younger children and guidance on how this can be done is provided in the information for each project.

Coverage of ICT National Curriculum Programmes of Study by each Project

Key Stage 1

Projects:	1	2	3	4	5	6	7	8	9	10
Finding things out										
1a. gather information from a variety of sources					✓					
1b. enter and store information in a variety of forms	✓	✓	✓							
1c. retrieve information that has been stored		✓	✓							
Developing ideas and making things happen										
2a. use text, tables, images and sound to develop their ideas		✓			✓					
2b. select from and add to information they have retrieved for particular purposes										
2c. plan and give instructions to make things happen	✓			✓						
2d. try things out and explore what happens in real and imaginary situations	✓		✓	✓						
Exchanging and sharing information										
3a. share their ideas by presenting information in a variety of forms		✓			✓					
3b. present their completed work effectively										

Key Stage 2

Projects:	1	2	3	4	5	6	7	8	9	10
Finding things out										
1a. talk about what information they need and how they can find and use it						✓				✓
1b. prepare information for development using ICT, including selecting suitable sources, finding information, classifying it and checking it for accuracy						✓				✓
1c. interpret information, to check it is relevant and reasonable and to think about what might happen if there were any errors or omissions					✓	✓				✓
Developing ideas and making things happen										
2a. develop and refine ideas by bringing together, organising and reorganising text, tables, images and sound as appropriate										
2b. create, test, improve and refine sequences of instructions to make things happen and to monitor events and respond to them				✓			✓		✓	
2c. use simulations and explore models in order to answer 'What if . . . ?' questions, to investigate and evaluate the effect of changing values and to identify patterns and relationships					✓		✓	✓	✓	
Exchanging and sharing information										
3a. share and exchange information in a variety of forms, including e-mail					✓	✓				
3b. be sensitive to the needs of the audience and think carefully about the content and quality when communicating information						✓				

Figure 1

Year groups covered by each project

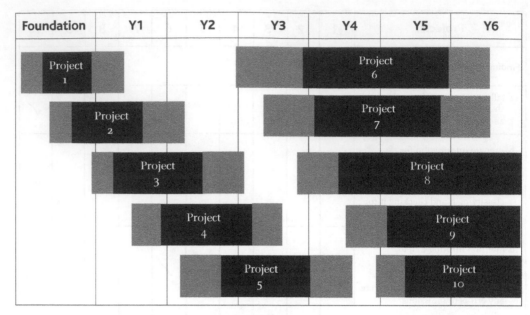

Figure 2

Links to the QCA scheme of work for ICT

Figure 3 indicates the relationship between the projects included in this book and the QCA ICT units in Key Stages 1 and 2. The key indicates which QCA units can be replaced, supported or extended by the projects in this book.

Links to the QCA Scheme of work for ICT in Key Stages 1 and 2

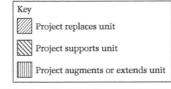

Key
- ▨ Project replaces unit
- ▧ Project supports unit
- ▥ Project augments or extends unit

Projects	1	2	3	4	5	6	7	8	9	10
Unit 1A: An introduction to modelling	replaces		supports							
Unit 1C: The information around us		supports								
Unit 1D: Labelling and classifying		replaces	augments							
Unit 1E: Representing information graphically: pictograms		replaces								
Unit 2D: Routes: controlling a floor turtle	supports			supports						
Unit 3A: Combining text and graphics						replaces				
Unit 3C: Introduction to databases						replaces				
Unit 3D: Exploring simulations					replaces					
Unit 4A: Writing for different audiences						augments				
Unit 4B: Developing images using repeating patterns					augments					
Unit 4D: Collecting and presenting information: questionnaires and pie charts						augments				
Unit 4E: Modelling effects on screen				augments			replaces			
Unit 5C: Evaluating information, checking accuracy and questioning plausibility										supports
Unit 5D: Introduction to spreadsheets								supports	replaces	
Unit 5E: Controlling devices								augments		
Unit 6A: Multimedia presentation										
Unit 6B: Spreadsheet modelling								augments	supports	
Unit 6C: Controlling and monitoring – What happens when . . .?								augments		
Unit 6D: Using the internet to search large databases and to interpret information										augments

Figure 3

Project Fact Card: Project 1: Using programmable toys

Who is it for?

- 4- to 6-year-olds (NC Levels 0–1)

What will the children do?

- Use a programmable toy to reinforce basic number skills and to develop vocabulary associated with 2D space

What should the children know already?

- That some toys can be programmed to carry out instructions

What do I need to know?

- How to operate a programmable toy with basic commands (forward, back, left, right, go, clear memory)
- How to change or charge the batteries in a programmable toy (if necessary)
- (Optional) How to change the step-size for some programmable toys

What resources will I need?

- A programmable toy
- A number line appropriate for the toy
- A maze or map appropriate for the toy

What will the children learn?

- That numbers can be associated with distances
- That small forward moves can be combined (i.e. the basis for addition)
- That backward moves negate forward moves (i.e. the basis for subtraction)
- (Some children) That four quarter turns are the same as one complete turn
- That programmable toys can be given instructions that they remember and which can be changed

How to challenge the more able

- Increase the distances to be programmed
- Provide more challenging environments for the children to explore

How to support the less able

- Provide more adult support
- Adjust the complexity of the activity to match the capabilities of the children

Why teach this?

- It introduces the children to ICT NC KS1 PoS statements 1b, 2c, 2d
- It acts as a precursor, complement or replacement for QCA ICT Scheme of Work Unit 1F and leads into Unit 2D
- It addresses Early Learning Goals PSE, CLL, MD, KUW
- It lays the foundations for Mathematics NC KS1 PoS statements Ma2, 1a, 1b, 1e, 1f, 2c, 3a; Ma3, 1a, 1d, 1e, 1f, 3a, 3b, 3c, 4a, 4b
- It reinforces NNS Reception Units Au/Sp/Su 1, 2, 4, 9

Using programmable toys

What will the children do?

Although the focus here is on the ICT aspects of the activities, it is important that the activities are integrated with ongoing work on developing basic number concepts.

Activity 1: Investigating the toy

This first activity enables the children to explore their own ideas about how the programmable toy works. Show the toy to the children and ask them to 'guess' what they think the buttons might do. Let the children offer opinions and then choose one to try his/her idea out.

Once the basic controls have been mastered, set a series of challenges – moving the toy forward and back:

⊙ Can you move the toy forward from here to here?

⊙ Can you move it forward to there in one step?

⊙ Can you move the toy back to where it started?

To enliven the activity, a context could be chosen (e.g. The Jolly Postman delivering mail to houses in a street, The Very Hungry Caterpillar crawling to particular fruits).

Once the children have grasped basic movements you could ask them to perform more complex moves:

⊙ Can you move the toy to there, pause, and then back where it started?

⊙ Can you move the toy forward to here, pause, then move on to there?

⊙ Can you move the toy to here, and then back to there?

If a mistake is made, the children should be asked what they need to change to put it right, rather than being told what they needed to do. It is important they appreciate they can use the toy to test their ideas and, if necessary, self-correct.

Activity 2: Moving along a number line

This activity is similar to Activity 1, except that a number line is laid out on the floor or table, but you ask the children to move from and to particular 'numbered' locations. For example:

⊙ Can you move from 0 to 4?

⊙ Can you move from 2 to 5?

⊙ Can you move from 4 to 1?

The activity could be contextualised by using a street and house numbers (e.g. for *The Jolly Postman*). Encourage the children to verbalise their predictions before testing them. For example, 'To move from 2 to 5 I am going to move forward 3 steps'.

Activity 2a (Extension to the above)

If some children master the above quickly and confidently, you could ask them to find different ways of moving to a particular target number in two steps. For example, to move from 0 to 5, one child might do this in a step of 3 followed by a step of 2; another child might do this as a step of 4 followed by a step of 1.

Activity 3: Exploring a map

Set out a simple maze (or plan) on a grid equivalent to the toy's step-size (most programmable toy suppliers provide ready-made grids and/or plans for their toys to explore). Ask the children to initially explain and then try out their ideas for moving the toy from one location to another. The level of difficulty can be adjusted to suit different children or the children's progress in solving the problems. For example:

⊙ A move involving no turns.

⊙ A move involving one (right angle) turn.

⊙ A move involving two turns.

⊙ A move involving a turn and a reverse.

What should the children know already?

Prior knowledge or experience is not essential for this activity. Most children will know about battery-operated toys and will be willing to explore how they work.

It is not essential that they are familiar with numbers as Activity 1 can be completed without the need for counting. By carrying out the activities, they will come to appreciate that being able to count and associate numbers with distances and the symbols on the number line is useful for helping with their predictions.

What resources will I need?

A programmable toy

There are currently four programmable toys appropriate for use in the Early Years or primary classroom:

⊙ *Roamer* – see http://www.valiant-technology.com/freebies/product1.htm

⊙ *Pip* and *Pixie* – see http://www.swallow.co.uk/

⊙ *Bee-Bot* – see http://bee-bot.co.uk/

They all perform a similar function by allowing the children to type in a series of instructions which the toy carries out. *Pixie* and *Bee-Bot* are small enough to be used on a table. The other two are larger but have the advantage that they include a number keypad and can be connected to a computer via a cable. Don't forget to purchase or recharge the batteries for your toy the night before you intend to use it.

A number line appropriate for the toy

A maze or map appropriate for the toy

The number line can be simply constructed from a roll of lining paper or redundant wallpaper. Alternatively, laminated number cards can be placed in a row at appropriate intervals. A visit to the relevant website for your programmable toy (see above) will allow you to buy additional resources such as these designed specifically for use with the toy.

What do I need to know?

How to operate a programmable toy with basic commands

Most programmable toys operate in similar ways. There is a set of buttons on the back of the toy which are pressed to instruct the toy. In most cases, the instructions are remembered by the toy in the sequence entered and then carried out when the **GO** or **START** button is pressed. With most of the toys, the sequence of instructions is held in the memory until it is cleared – any new instructions are added to the end of the previous list. Thus a complex problem can be solved in stages, provided the toy is returned to its starting position after each run.

Some toys enable the last instruction to be cleared with a single press of the **CLEAR** button. This is very useful if the children are trying to solve a complex problem and need to change the last instruction entered. It would be worth while experimenting with the toy or checking the instructions before using it with the children to familiarise yourself with the way the toy's memory is used.

How to change or charge the batteries in a programmable toy

Programmable toys tend to drain their batteries quite quickly as power is needed for driving the motors as well as powering the electronic memory. The toys can be powered by disposable or rechargeable batteries, the latter being the most economical. It is wise, before setting up a session with a programmable toy, to ensure there is a supply of spare, fully charged batteries available as it can be frustrating for the children if the toy runs out of power midway through their activity.

(Optional) How to change the step-size for some programmable toys

The larger programmable toys (e.g. *Roamer, Pip*) enable the teacher to change the size of the step the toy moves forward or turns.

Roamer and *Pip* can have their step-size altered by the teacher

For example, **FORWARD 1** usually moves the toy forward by its own length. This is useful if there is plenty of space or if the toy is exploring a grid divided into squares equivalent to its own size. However, if space is limited or a group of children is ready to move on to exploring larger numbers, then the step-size can be altered. So, for example, **FORWARD 1** might move the toy forward one centimetre. Similarly, some toys use one degree as the default turn-size. You may wish to change this so that **RIGHT 1** turns the toy through a right angle (i.e. 90°). You will have to consult the manual which accompanied the toy to find out how (or if) the step-size can be altered.

What will the children learn?

You will notice that the majority of the learning outcomes for this project are mathematical. The ICT learning outcome should also be stressed as this lays the foundation for later work using floor turtles and LOGO programming.

That numbers can be associated with distances

ICT can be very effective in bridging the gap between concrete (enactive) experience and abstract (symbolic) notation. Through making predictions about the distance which needs to be moved and then watching the toy confirm or refute their suppositions, the children are learning to associate numbers with quantities in real life. For some, visualising the toy moving along a number line will prove to be a lasting mental image which will enable them to understand more complex concepts.

That small forward moves can be combined (i.e. the basis for addition)

That backward moves negate forward moves (i.e. the basis for subtraction)

If children miscalculate a distance, you can either suggest they start again or, once they have understood the relationship between the distance and the numbers represented, then they could be asked if they can work out how many more the toy should move forward (or back) to reach the target. This can then be followed by repeating the same total move as a single step (or as a combination of two or more different steps – see Activity 2a above).

(Some children) That four quarter turns are the same as one complete turn

As an extension to Activity 3, some children could investigate larger turns than a quarter turn (e.g. **RIGHT 1**). For example, if a journey can be completed by moving **FORWARD 2, RIGHT 1, FORWARD 3**, the children could be asked if the same journey could be completed by turning LEFT (i.e. **FORWARD 2, LEFT 3, FORWARD 3**). Some children might also discover that **FORWARD 2, BACK 2** can be achieved by **FORWARD 2, RIGHT 2, FORWARD 2**.

Experimenting with quarter turns lays foundations for later work on angular measurement.

That programmable toys can be given instructions that they remember and which can be changed

The children will quickly realise that the toy can remember quite long lists of instructions (for example, the tiny *Bee-Bot* can store 40 instructions). Emphasising the toy's memory capacity from time to time and asking them how they could record the instructions they will give to the toy (e.g. on paper) provides a useful link from first-hand (enactive) experience through the use of pictures (iconic representation) which could ultimately lead to the use of symbols (numerals). For example, **FORWARD 3** could be represented as:

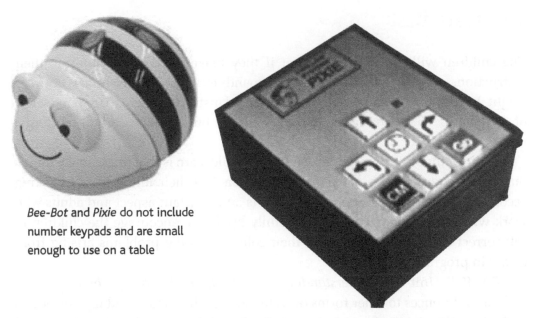

Bee-Bot and Pixie do not include number keypads and are small enough to use on a table

The complexity of the toy's keyboard will determine the transition from one representation to the other. *Bee-Bot* and *Pixie* do not include a number keypad and hence are easier to use, whereas *Pip* and *Roamer* have numerical keypads.

Challenging the more able and supporting the less able: modifying the project for older and younger pupils

Altering the distances to be programmed

As children become more confident with the toy they should be able to cope with larger numbers. You may need to change the step-size for the toy if the distances become very large (see above).

Varying the complexity of the environments the children explore

With Activity 3, the complexity of the mazes and maps which the children explore could be varied to match their capabilities. Those who are more experienced or confident could be given quite complex scenarios to explore, while those lacking in confidence could be given relatively straightforward maps.

Some children might design and produce their own mazes or maps, using a prepared grid as the basis.

Varying the level of support provided

Initially, all children will require a high level of support and carefully structured activities while they explore how the toy works. The level of challenge and the speed with which the children move through the tasks can be adjusted by a supervising adult to meet the needs of the children.

As the children become more familiar with the toy, they might suggest their own activities. A range of ideas for related activities and some resources are provided on the websites for the various toys (see *What resources will I need?*).

Why teach this?

The children will quickly realise that if they make mistakes in entering their instructions, the toy will carry out the commands exactly – the machine is not very bright. In addition, the toy is able to store several instructions and carry out a series of commands – i.e. it has a memory just like us; but unlike us it will not forget unless it is told to.

The most important concept the children should learn is that they can see immediately the consequences of their actions and adjust their initial ideas by taking account of the feedback they receive. This is an important aspect and adults who work with the children should be carefully briefed to encourage the children to self-correct rather than suggesting their solutions to the problems arising from errors in programming.

QCA ICT Unit 1F: *Understanding instructions and making things happen* includes references to other forms of technology such as video and tape recorders and relates the giving of instructions to physical activities such as moving around the school. These activities complement those in this project and hence could be easily combined.

The activities in ICT Unit 1F are wider ranging and can be integrated into a number of different contexts. The activities in this project are more focused and lead more specifically into ICT Unit 2D: *Routes: controlling a floor turtle.*

The children should be encouraged to share their ideas and take account of others' suggestions when solving the problems. Expressing their ideas clearly and precisely will contribute to their linguistic development. The activities provide meaningful opportunities for the children to develop basic number skills by linking physical action with symbolic notation. Programmable toys are highlighted specifically in the Early Learning Goals in the ICT strand for 'Knowledge and understanding of the world'.

The programmable toy enables the children to associate movements with steps along a number line and hence to associate physical actions with numerals and direction with number operations (**FORWARD** with addition and **BACK** with subtraction). For some children the visual images thus created will provide them with valuable internal representations for the mental manipulation of number problems.

Encouraging the correct use of vocabulary to explain their ideas and to aid accuracy in the communication of their predictions helps to emphasise that mathematics requires the precise use of language.

Similarly, the exploration of physical space vicariously through controlling the toy encourages the children to be precise in their use of language by helping them differentiate between right and left, the amount of turn, and associating numbers with distances to be travelled.

See also *Arts* Project 8 (*LOGO animation*) and *Arts* Project 9 (*Controlling external devices*) for related activities.

Project Fact Card: Project 2: Counting

Who is it for?

- 4- to 6-year-olds (NC Levels 0–2)

What will the children do?

- Following a non-computer-based activity in which the children sort cuddly toys and rearrange them into columns, the children represent the same activity on computer as a pictogram and a column graph. They then use this information to answer questions and are encouraged to suggest how they might use this approach to answer other similar questions

What should the children know already?

- Nothing, though experience of sorting other sets of objects could be a useful precursor

What do I need to know?

- How to use 2Simple's *2count* program (or similar such as BlackCat *Counting Pictures 3*)
- How to create a counting file for *2count* (or similar) using Clip Art images

What resources will I need?

- Cuddly toys for sorting into sets (or other objects related to an ongoing class topic)
- A counting or simple graphing program such as BlackCat *Counting Pictures 3*
- Some pre-prepared graphs for interpretation
- (Optional) Clip Art images to enable you to create your own counting files

What will the children learn?

- That objects can be classified and sorted into sets
- That sets can be compared and information about these comparisons can be communicated through such vocabulary as: the same as, more than, less than, bigger than, smaller than, biggest, most, smallest, least
- That objects can be represented by images, icons or blocks from which the same information can be deduced
- Some children will also recognise that numbers and numerals can be used to accurately communicate information about the size and differences between sets of objects

How to challenge the more able

- Ask more challenging questions
- Repeat the activity but with a larger range of objects
- Ask the children to devise their own related activity and work more independently as a group

How to support the less able

- Pitch questions which are matched to the children's understanding
- Repeat the activity but with other and fewer objects for sorting and comparing
- Provide more adult support

Why teach this?

- It introduces the children to ICT NC KS1 PoS statements 1b, 1c, 2a, 3a
- It complements QCA ICT Scheme of Work Unit 1C and/or replaces Unit 1E
- It addresses Early Learning Goals PSE, CLL, MD, KUW
- It lays the foundations for Mathematics NC KS1 PoS statements Ma2, 1a–f, 2a, 2c
- It reinforces NNS Reception Units Au/Sp/Su 1, 2, 7

Counting

What will the children do?

Activity 1: Sorting toys

This is a preparatory non-computer activity. Present a collection of toys to the children and ask them to take turns to sort them into different sets based on their own criteria (e.g. size, animal, colour, etc.). The other children (and you) have to try and work out what criteria the child is using. The activity could be enhanced through the use of hoops or sets made from skipping ropes on the floor. An empty set (e.g. blue toys) could be introduced.

Once the toys have been sorted, ask questions about the sets before the next child sorts them. For example, 'Which set has the most toys in it?' 'Which set has the least?' 'Are there any sets with the same number of toys in them?', etc.

After a while, arrange the sets into 'columns' to help with the comparisons and also to introduce the concept of a 'real' pictogram in preparation for the next activity.

At this stage it is not essential to count the number of items in each set in order to make the comparisons; however, some children will naturally count the toys. This may or may not help in making comparisons.

Activity 2: Using the computer to count and graph the toys

Before this activity, you need to choose a BlackCat *Counting Pictures 3* (or similar) file (see *What do I need to know?*, below). In this example we have used the Favourite Toy file.

Sort the toys physically into sets (e.g. by type or colour) then arrange each set into columns to form a 'live' pictogram. Ask a child to count or touch each toy in one set and at the same time add a picture to the relevant column in *Counting Pictures 3*. Ask the children questions about the relative sizes of the sets, using the information in the chart and/or the 'live' pictogram.

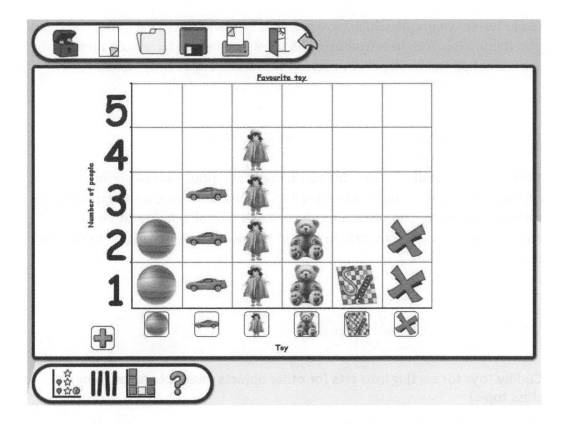

If *Pictogram* (Kudlian), *Starting Graph* (RM) or *Counting Pictures 3* (BlackCat) is being used, the pictogram can be changed into a block graph and a bar chart. This feature is particularly useful in helping the children associate the abstract blocks with the objects they represent. Careful questioning should be used to help the children make this important association.

The relationship between the images on the screen and the sets of toys on the floor or table should be emphasised throughout the activity to help the children appreciate the connection between the computer representation and the actual toys. Only move on to the block graph if you feel the children have fully grasped this relationship.

Activity 3: Sorting and recording another set of items

Ask the children to decide how they might sort, record and compare another set of objects (e.g. fruit, books or clothes). The choice of objects could be decided by checking the in-built pictogram files or by creating another set with Clip Art images. Repeat Activities 1 and 2 with the other set of objects but with the children taking the lead on suggesting what needs to be done next. You could use a puppet or one of the toys which needs to be shown what to do.

Activity 4: Consolidation and moving on

Depending on the learning needs of the children in the group, you could repeat the above activities to consolidate the relationship between the images, blocks or bars and the objects they represent. Those children who are ready to move forward might be shown graphs which have been made previously to see if they can interpret them. These can be introduced as a set of graphs that have been prepared by some children in another class or school which need to be interpreted.

What should the children know already?

There is no necessity for the children to have any prior experience of counting, sorting or using the computer to record their work. However, some experience of sorting, classifying and discussing the relative size of sets would be helpful to ensure that the initial activity is firmly understood before moving on to the more demanding concepts.

What resources will I need?

Cuddly toys for sorting into sets (or other objects related to an ongoing class topic)

The toys could be the children's own as part of a class topic or, if available, a set provided by the school. Alternatively, the tasks could be modified using any sets of objects related to the class topic (e.g. fruits, foods, pictures of animals, beans, etc.). The number of objects in each set and the number of sets will be dependent on your assessment of the children's capabilities and readiness for making numerical comparisons. You should also consider whether it is appropriate to include an empty set.

A counting or simple graphing program

2Simple *2count*, RM *Starting Graph*, Kudlian *Pictogram* and BlackCat *Counting Pictures 3* will all allow you to create your own picture files. Although this is not essential, it will enhance the children's learning if you are able to create categories which relate to the toys you are sorting. For example, if you are using toys which the children have brought in, you may need to create a file which includes the types of animal (e.g. bear, cat, rabbit, seal, hedgehog). This file may not exist as one of those provided with the software.

Some pre-prepared graphs for interpretation

A series of two or three graphs could be prepared for use with children who are ready to move forward. These graphs (pictograms and/or block or bar charts) could depict objects similar to those used in the early activities or could be quite different (e.g. methods of transportation to school).

(Optional) Clip Art images

These will be useful if you want to create your own files for sorting. These can either be purchased quite cheaply on CD-ROM or can be downloaded free of charge from various websites. Alternatives are to create your own using a painting program, scan in children's drawings or use digital photos.

What do I need to know?

How to use 2Simple's *2count* program (or similar – e.g. RM *Starting Graph*, Kudlian *Pictogram*, BlackCat *Counting Pictures 3*)

As the name suggests, 2Simple software's *2count* program is very easy to use. Once it is loaded (**Start menu > programs**), clicking on the **New** graph icon in the top left corner of the window will present you with the set of built-in pictograms. Select the one which you feel is most appropriate and click

the **Yes** button. Once the empty graph appears, clicking on the button at the base of a column will add a new picture.

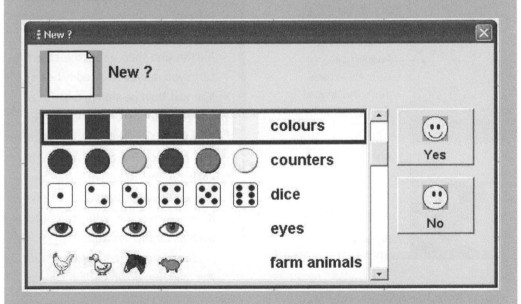

How to create a counting file for *2count* (or similar) using Clip Art images

Finding Clip Art images
If you are going to use Clip Art images, you will need to check you have the right ones to hand before creating your own *2count* file. If these are on a CD-ROM, it is advisable

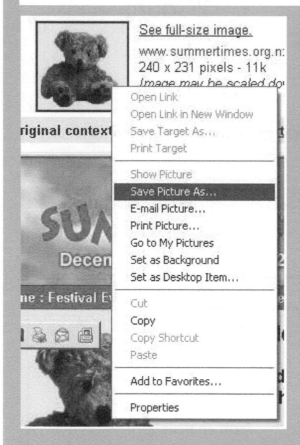

to browse through them first and make a note of the names of the images and the folders in which they can be found.

You can search the internet for suitable Clip Art. Search engines such as Google or AltaVista enable you to search specifically for images.

Click on the **Image** search button and then enter the search term you require (e.g. teddy bears). You will then be shown a series of images. Clicking on the image will take you to the website containing the picture where you can right click on the required image and choose **Save Picture As . . .** to save it somewhere on your computer such as on the desktop.

Some useful educational Clip Art can be downloaded free of charge from the Softease website (http://www.softease.com/clipart.htm).

If you have the earlier version of *2count*, you will have to go to the 2Simple website (www.2simple.com) and download a small program called *2count Picture Editor*. Follow the instructions for using this carefully to create your own files.

The more recent version of *2count* makes the process of adding your own files more straightforward. Click on the **New** graph button (as above) and then highlight the first item (**My Pictures**), then click the **Yes** button:

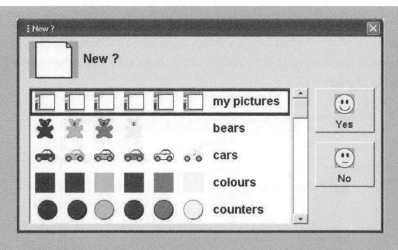

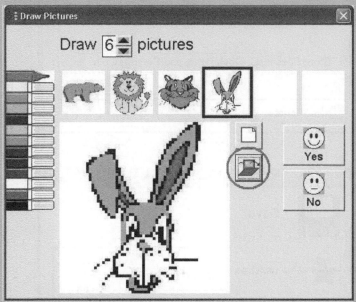

Click on the first box and then click on the button which resembles an **open folder.** Locate the first of the images you want to use (i.e. on the CD-ROM or saved on your desktop) and click the **Open** button.

Repeat the process until you have imported all the images needed for your sorting activity. Then click on the **Yes** button. At some point you need to save the file you have created. Open the *Teacher controls* by holding down the **Ctrl** (Control) and **SHIFT** (⇧) keys and typing the letter O.

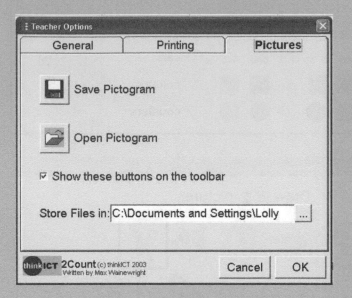

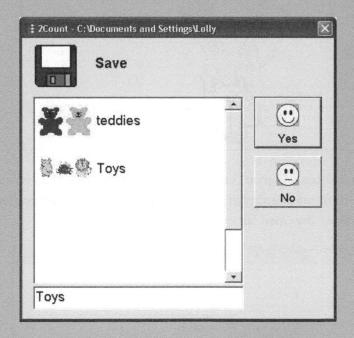

Click on the **Pictures** tab. Click on **Save pictogram** and then give your file a suitable name and click the **Yes** button. To load your saved file, follow the same procedure but click on the **Open Pictogram** button.

NOTE: Some images cannot be opened by *2count* or the other pictogram programs mentioned. Most will open images which have been saved in bitmap (.bmp) or JPEG (.jpg) formats. If you find that an image cannot be opened (usually it is not shown when you open the relevant folder), its format can be changed by opening the image in *Paint* (**Start menu** > **Programs** > **Accessories** > **Paint**) and then saved as a bitmap or JPEG image by selecting **Save As ...** from the **File** menu and then clicking on **24 bit-BITMAP** or **JPEG** on the **Save as type** pull-down menu. You then give the image a suitable name and click on the **Save** button.

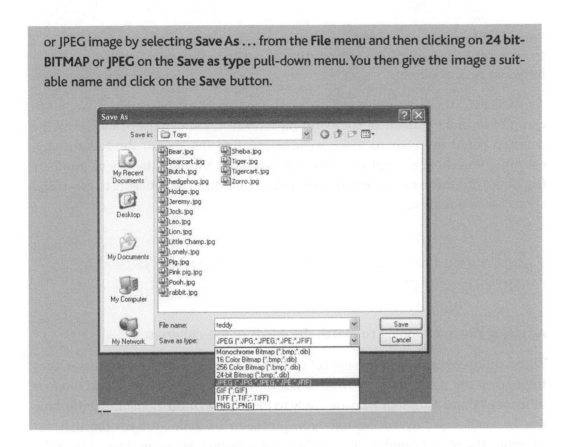

What will the children learn?

That objects can be classified and sorted into sets

That sets can be compared and information about these comparisons can be communicated through such vocabulary as: the same as, more than, less than, bigger than, smaller than, biggest, most, smallest, least

This is a fundamental concept and skill which forms the basis for much work in mathematics and science. These activities contribute to the children's development by providing contexts in which they can identify differences and similarities based on defined criteria and then to articulate through responding to your questioning.

That objects can be represented by images, icons or blocks from which the same information can be deduced

Being able to change the way the information is presented on screen dynamically helps the children develop their understanding of the relationships between the real objects and, ultimately, abstract symbols (i.e. numerals) which represent those objects. Building children's ability to visualise by associating the abstract and the concrete is something which computer programs used in this way can achieve well, provided the activities are scaffolded to help the children form these associations. The emphasis is therefore on questioning rather than explanation and instruction.

Some children will also recognise that numbers and numerals can be used to accurately communicate information about the size and differences between sets of objects

As indicated, these activities can be completed without the children having to count or to make numerical comparisons. However, some, most or maybe all the children could be ready to more accurately define the size of the sets through counting (i.e. there are eight teddies). Comparisons through the use of numbers (e.g. there are three more teddies than rabbits) are difficult concepts to grasp and should not be rushed.

Challenging the more able and supporting the less able: modifying the project for older and younger pupils

Adapting the questioning to match the needs of the children

Those children needing more support will need the pace to be eased to ensure they fully grasp the relationship between what they are seeing on screen and the toys in front of them. Children who are ready to move on to more demanding thinking can be challenged by asking for more precision in their answers through, for example, the use of numbers (e.g. 'Which set/column has one more in it than the lions?').

Adapting the activitiesz

Children requiring more support will need to have the concepts consolidated through repetition of similar activities as in Activity 3. The number of items could be reduced or increased as could the range of possible criteria for sorting. For example, objects could be sorted on secondary criteria such as the material from which they are constructed. Those who are capable of more challenging work can take more responsibility for guiding the activities or could move on to the extension to Activity 4.

Adjusting the level of support

Those requiring more assistance may need to have adult input throughout all activities whereas those who are more confident and capable could be briefed initially and work independently on a sorting and recording activity of their own.

Why teach this?

For some children, the information which computers (and TV) present is separate from their own experience. These activities help them to appreciate that computers can be used to present and manipulate information which is relevant to them. Knowing that information can be entered, changed, stored and accessed later – and that the information which others have entered can be accessed and manipulated – is an important stage in children's learning.

QCA ICT Unit 1E: *Representing information graphically: pictograms* focuses primarily on ICT aspects without embedding them in a mathematical or other subject-related context. This project aims to show how the same ICT objectives can be achieved while building important mathematical understanding.

ICT Unit 1C: *The information around us* concentrates on text and images as sources of information. Pictograms convey mathematical information and relationships visually and hence this project could be integrated or cross-referenced with Unit 1C.

By working in groups and having to take account of others' views and ideas, the children will be learning to co-operate and share. Careful questioning will enable the children to frame their responses and by probing elaborate on the reasoning behind their criteria for sorting, for example. Accurate use of vocabulary in describing the sets and in making comparisons contributes to mathematical development as does manipulating the toys and other objects and learning to associate the objects with images, blocks, columns and numerals. Learning that objects can be classified and generalisations made about categories is a key concept for helping children understand the world, as is recognition that computers can help them record their own experiences and interpret what 'others' have done.

Making comparisons and eventually quantifying relationships between sets (i.e. more than, three less than) forms the basis for calculation, particularly subtraction (difference) and addition (more than). Bridging activities which enable the children to associate visual representations (pictures, icons, blocks, columns) and abstract notation (numerals) with reality help to build children's internal models which can be used to help them manipulate mathematical information through internal visualisation. These activities make effective use of technology to help in this process.

The computer is being used as a way of recording the children's experience in sorting the objects and then making comparisons. Charts and graphs are highly effective ways of enabling us to make sense of numerical information. Consider how much easier it is to read a graph portraying complicated numerical information than it is to try and make sense of the numbers themselves. Helping children appreciate the way pictograms and block graphs can show information in the real world and can help them answer questions is an important foundation concept.

See also *Science* Project 5 (*Graphical representation of data*) for related activities.

Project Fact Card: Project 3: Shopping

Who is it for?

- 4- to 6-year-olds (NC Levels 0–2)

What will the children do?

- The children will make productive use of a computer as part of a play environment – in this example the context is a supermarket. 'Shoppers' take their items to the check-out where the shopkeeper enters the items and takes the money. As a follow-up some children can shop in a virtual supermarket

What do I need to know?

- How to install software from CD-ROM
- (Optional) How to attach the *Hasbro Playskool Store PlaySet* 'till' to the keyboard
- How to install and use the *Going Shopping* program
- How to configure SPA software's *Let's Go Shopping* software to match the capabilities of your children
- (Optional) How to use the *Playskool Store* software

What should the children know already?

- That computers are used in a range of different ways in real contexts (e.g. to book a holiday, at the check-out in a supermarket, to find a book in the library, etc.)
- How to read numerals (the range can be adjusted to suit the capabilities of the children)
- How to manipulate the mouse to point and click

What resources will I need?

- *Going Shopping* program
- (Optional) *Hasbro Playskool Store PlaySet*
- SPA software's *Let's Go Shopping* or similar
- Shop role-play area with counter, pretend stock and display

What will the children learn?

- That supermarket tills are a form of computer
- How to relate real objects to images on screen
- How objects are classified
- How to interpret amounts of money displayed on screen

How to challenge the more able

- Increase the upper limit of the item costs in the *Let's Go Shopping* simulation program
- Use an alternative shopping program or an online shopping game such as *Loose Change*

How to support the less able

- Lower the upper range of the prices in *Let's Go Shopping*
- Provide more adult support

Why teach this?

- It introduces the children to ICT NC KS1 PoS statements 1b, 1c, 2d
- It complements QCA ICT Scheme of Work Unit 1A and/or augments Unit 1D
- It addresses Early Learning Goals PSE, CLL, MD, KUW
- It lays the foundations for Mathematics NC KS1 PoS statements Ma2, 1a, 1b, 1e, 1f, 2a, 2c
- It reinforces NNS Reception Units Au/Sp/Su 1, 2, 5, 11

Shopping

What will the children do?

Activity 1: Familiarisation with the *Going Shopping* program and (if available) *Playskool Store* 'till'

Once the equipment has been set up (computer, program, *Playskool Store* 'till' attached to the keyboard, some 'stock' from the supermarket role-play area), gather the children around and ask about their experiences of shopping. Focus particularly on how different items are located in different parts of the supermarket and how the program reflects this. Demonstrate also the check-out feature, and how the 'till' knows how much everything costs and can add them up automatically.

Select a child to be the check-out operator and another to choose an item and take it to the 'till'. Show the check-out assistant how to locate and 'check-out' the item (with guidance from the other children if necessary). Once the price of the item is displayed, the shopper is asked for payment with play money.

Repeat with other 'volunteers' until the children seem to be familiar with the equipment.

Activity 2: Role play

Set up the equipment and computer in the shop role-play area and organise the children into groups. (NOTE: The items in the shop should match the items in the program – 'departments' can be turned off using the teacher controls.)

Activity 3: Familiarisation with the *Let's Go Shopping* or *Playskool Store* simulation

Set up a computer with the software. If *Let's Go Shopping* is being used, configure it for whichever scenario is appropriate for the children (see *What do I need to know?*, below). Gather the children around and ask them about the *Going Shopping* program and/or the *Playskool Store* 'till' and how this activity compared with shopping in a real supermarket.

Demonstrate the software, and ask the children to take on the different roles and suggest what needs to be done next.

Activity 4: *Let's Go Shopping* or *Playskool Store* simulation

The software is installed on a classroom computer and the children are organised into groups to use it. If appropriate, tell each group to explain to the next group how the software works.

What should the children know already?

That computers are used in a range of different ways in real contexts (e.g. to book a holiday, at the check-out in a supermarket, to find a book in the library, etc.)

Through discussion children's attention should be drawn to the different places they have seen computers used in the world around them. Sometimes the application may not obviously resemble a computer in its conventional sense (e.g. an ATM machine outside a bank, a ticket machine on a bus or tram) and children should come to appreciate that computers are used in a range of situations where information needs to be stored and manipulated.

How to read numerals (the range can be adjusted to suit the capabilities of the children)

Although it is not essential that numerals are read accurately for the first two activities, the success of the third and final activities relies on the children correctly identifying the amounts shown on screen. The first two activities could be used for familiarisation with and reinforcement of numeral recognition but will need to be combined with a range of other activities.

How to manipulate the mouse to point and click

The activities can be used to reinforce mouse skills but could prove frustrating if a child lacks the control needed to select the relevant items on screen.

What resources will I need?

Going Shopping program

This is available on the CD-ROM which comes with this book.

(Optional) *Hasbro Playskool Store PlaySet*

Although the activity can be planned without this, the addition of the 'till' to the computer keyboard makes the use of the resource highly motivating.

SPA software's *Let's Go Shopping*

Other shopping programs can be used, but *Let's Go Shopping* provides a realistic and highly configurable environment appropriate for moving the children on from the initial role-play activity.

Information about the program can be found at http://www.spasoft.co.uk/letsgo.html

Shop role-play area with counter, pretend stock and display

The extent to which you replicate a supermarket depends on the space, resources and time you have available to set it up.

What do I need to know?

How to install software from CD-ROM

Software is usually supplied with instructions, including how it should be installed. In most cases, inserting the CD into the CD drive will result in a series of on-screen menus explaining how to complete the installation. Sometimes, however, your computer may not be set up to run CDs automatically in which case you will have to start the CD-ROM yourself. Provided you can gain access to the computer's desktop the following procedure should be followed:

⊙ Insert the CD-ROM into the CD drive and close it.
⊙ On the desktop, double-click on the **My Computer** icon.

⊙ From the list of drives shown, double click on the CD drive (it might be listed as Drive D on your computer).

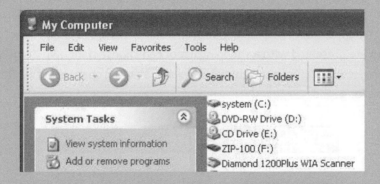

⊙ The CD-ROM installation disc should now load and present you with the instructions for installing the program. If not, double click on the **Setup** file.

(Optional) How to attach the *Hasbro Playskool Store PlaySet* 'till' to the keyboard

This is a relatively straightforward process and is fully explained in the instructions which accompany the *PlaySet* equipment. If the instructions have been misplaced, the most important thing to remember is to ensure that the till unit is situated accurately over the keyboard (so the key in the upper left corner of the keyboard sits snugly in the relevant place under the till).

How to install and use the *Going Shopping* program

⊙ Go to the desktop (by closing all other windows) and double click on **My Computer**.
⊙ Double click on the icon for the CD-ROM drive.

- ⊙ Double click on the **Project 3** folder.
- ⊙ Double click on the **Resources** folder.
- ⊙ Double click on the **Shopping** folder.
- ⊙ Double click on the **Shopping.exe** file to install the program.

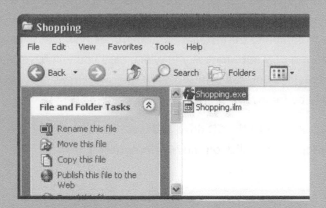

How to configure SPA software's *Let's Go Shopping* software

To match the software to the capabilities of your children, you will need to configure the program before each group uses it. The instruction manual which accompanies the software is very comprehensive, but if it has been misplaced, the teacher controls can be accessed by clicking on the menus on the first screen of the program. These allow you to set the upper limit for spending, the units the prices are rounded up to, whether the children are presented with a shopping list or can choose their own items, whether the children need to present the right money or will be given change and whether there is music and/or sound.

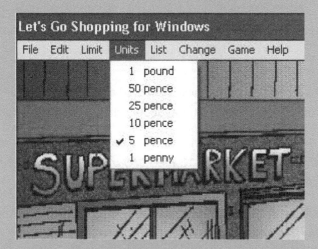

(Optional) How to use the *Playskool Store* program

To decide which parts of the *Playskool Store* program you decide to use with the children it is advisable to work through it beforehand. The scenarios presented in the various departments can be confusing if the children are not carefully briefed. The practice games in each department are more straightforward but have limited educational potential.

If you do decide to use this software, it is advisable to have adult supervision.

What will the children learn?

That supermarket tills are a form of computer

Because the *Playskool Store* 'till' is attached to their classroom computer the children will quickly appreciate that it is performing a similar process to the check-out till in the supermarket. It should be explained to them that the supermarkets buy special computer tills that hide all the familiar features of a computer. An organised trip to a local supermarket will help the children recognise the similarities between their role playing and the real thing.

How to relate real objects to images on screen

The images shown on the computer screen for both programs will resemble the objects the children will have in their role-play area but clearly not be direct representations. Relating real objects to their iconic representations on screen is an important foundation concept for both ICT and mathematical development.

How objects are classified

Both the *Going Shopping* and *Let's Go Shopping* programs require the children to classify the items by looking for them in the right sections of the supermarket. Recognising that objects in the world can be organised according to similarities and differences is fundamental to scientific and mathematical understanding.

How to interpret amounts of money displayed on screen

It is not essential that the children are able to read the amounts shown on the screen in the *Going Shopping* program, which is why it is desirable to follow this activity with either the *Let's Go Shopping* or the *Playskool Store* activities. These rely on the children being able to interpret the prices and, with the more advanced settings, offer the correct money. The free online *Loose Change* program (see below) further enhances this aspect of learning by requiring the children to select the right coins to 'pay' for items shown on the screen.

Challenging the more able and supporting the less able: modifying the project for older and younger pupils

Modifying the range of costs

The *Let's Go Shopping* program has a range of teacher controls which enable you to tailor what it presents the children.

Using other shopping software or an online shopping simulation

A wide range of shopping programs is available; however, very few of them are suitable for very young children. The following could be used to provide variety or additional challenge:

The Oxford Primary FunZone

This website provides a free shopping game called *Loose Change* which requires the children to pay for items by dragging the requisite coins from their purse:

http://www.oup.co.uk/oxed/primary/funzone/funzone.html/omzshopping.html/

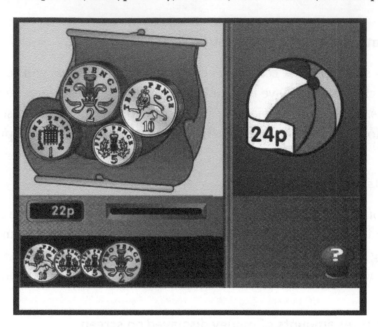

Percy's Money Box *(Neptune Computer Technology)*

This website includes several activities designed to help with coin recognition, paying for items, calculating change and recognising equivalent configurations of coins:

http://www.neptunect.co.uk/products/percysmoneybox/introduction.htm

LifeSkills – Time and Money *(Learning and Teaching Scotland)*

The program includes shopping activities set in a supermarket and a café and requires the children to calculate the change required. It is intended for older (special needs) children but could be used with younger children requiring more challenge.

Using the i-board shopping activities with an interactive whiteboard

I-board demonstration activities are available free of charge at:

http://demo.iboard.co.uk/screens/thread_home.htm?thread_id=9#

This link includes five different interactive on-screen activities associated with money and shopping.

Toy Shop – *an interactive game for two players*

Distributed to all schools by the Department for Education and Skills (DfES), free of charge in the Mathematics and ICT pack, *Toy Shop* presents the children with a series of items at various prices for which the players take turns to pay a coin. The child who pays the last coin 'buys' the object and wins the game. The children need to develop a strategy for ensuring they are the last to pay for the item.

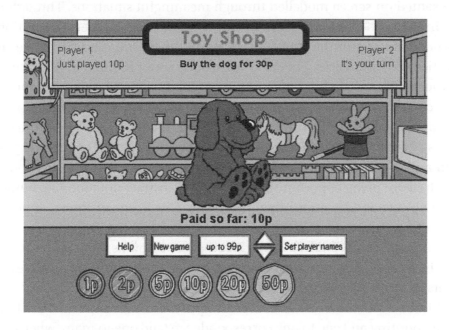

It can also be downloaded from the Primary National Strategy website:

http://www.standards.dfes.gov.uk/primary/publications/mathematics/12896/

Using adult support

Children lacking in experience or confidence may require additional support from an adult, particularly when using the *Playskool Store* program.

Why teach this?

Both programs model the way in which computers are used in the real world to store information about objects. The children are using rudimentary databases to access this information and view the results. This lays the foundation for searching for particular data by the use of categories (i.e. narrowing down a search). They are also deepening their awareness of ways in which ICT is used to make life easier.

QCA ICT Unit 1A: *An introduction to modelling* requires the children to explore an imaginary world through an adventure program or model the real world through a simulation. Both these activities simulate or model shopping scenarios and hence can be used as a direct replacement for some of the activities in Unit 1A.

In addition, ICT Unit 1D: *Labelling and classifying* suggests the children sort and describe objects. Sorting and classifying the items in the shop and relating them to the 'departments' in both shopping programs can be used to lead into, support or replace these activities. A word bank, such as that suggested in Unit 1D, could be used in conjunction with this project to encourage the children to record their experiences – this satisfying many of the learning outcomes from both units of study.

The role-play activity provides opportunities for children to enhance their communication skills through interacting with others in an everyday context. Mathematical development is supported as they encounter numbers and quantities presented on screen modelled through meaningful situations. Through identifying in which 'department' an item will be found, the children are reinforcing their understanding of the way objects around them are classified in terms of similarities and differences. They also are making use of ICT in real-world contexts and through structured adult intervention can reinforce the development of appropriate technical vocabulary (e.g. mouse, click, pointer, screen, icon, etc.).

By encountering the relationship between numbers and quantities (e.g. 4 packets of crisps) and between quantities and costs (1 packet of crisps is 5p), initially in a play context and then later in a more structured virtual environment, the children are given opportunities to associate abstract notation (i.e. numerals) with the objects and situations they represent. This is one of the most powerful features of mathematics – i.e. the structured representation of physical actions (or operations). Those children who can read the amounts shown on the screen will begin to appreciate the relationship between quantities and costs (e.g. that 2 packets of sweets costing 5p will always cost 10p).

In addition to the above, the children are provided with opportunities to practise their counting and one-to-one correspondence (and one-to-many when considering costs). They are also encountering the concept of addition of single items and repeated addition in the calculation of costs. The shopping program presents them with problems which require decisions on the most appropriate solution.

Project Fact Card: Project 4: Exploring with directions

Who is it for?

- 7- to 8-year-olds (NC Levels 2–3)

What will the children do?

- After an initial recap on using programmable toys, the children will explore some LOGO screen 'worlds' using simplified LOGO controls (similar to a programmable toy). They will then solve various maze problems by applying the knowledge gained in the introductory activities. Some children may move on to designing their own mazes for others to solve

What should the children know already?

- How to control a programmable toy (see Project 1)

What do I need to know?

- How to access and install web-based activities
- How to use a simple LOGO turtle graphics program
- How to install programs from CD-ROM
- (Optional) How to create additional mazes for a turtle graphics program using a *Paint* program

What resources will I need?

- A programmable toy
- A simple turtle graphics program
- LOGO – any version will be suitable
- *Crash* – provided on the CD-ROM accompanying this book

What will the children learn?

- About body-centred geometry (i.e. left, right, forward, back)
- That regular shapes can be drawn using a repeated series of the same instructions
- And hence, the properties of some familiar regular shapes
- The foundations of LOGO turtle graphics programming
- That problems can be solved in different ways
- That LOGO procedures comprise a series of instructions which can be amended

How to challenge the more able

- Explore more complex screens using more advanced controls
- Create mazes for others to solve

How to support the less able

- Allow more exploration time with the basic turtle programs
- Create simpler mazes for the children to solve
- More adult support and/or peer tutoring

Why teach this?

- It covers ICT NC KS1 PoS statements Exents 2c, 2d; KS2 PoS statement 2b
- It complements QCA ICT Scheme of Work Unit 2D and augments or replaces Unit 4E
- It addresses Mathematics NC KS1 PoS statements Ma2, 1a–c; Ma3, 1a, 1d, 2a, 3a–c, 4a, 4b; KS2 PoS statements Ma2, 1a, 1k; Ma3, 1c, 1g, 2a, 2b, 3a
- It reinforces NNS Y2 Units Au/Sp/Su 5–6; Y3 Units Au/Sp/Su 4–6

Exploring with directions

What will the children do?

Activity 1: Demonstration with programmable toy

Set up the programmable toy on a table or some floor space with the children gathered around. Ask the children what they can tell you about the toy and how it can be programmed. Set the children a series of challenges with increasing levels of complexity to gauge their knowledge and understanding:

⊙ How would you program the toy to travel in a straight line from here to knock over this skittle/can/bottle?

⊙ As above but also return to knock over another skittle placed behind its starting point.

⊙ Knock over two skittles requiring a turn of 90°.

⊙ Knock over skittles requiring two turns of angles other than 90°.

⊙ Make sure the children understand the basic directional controls and how the memory stores a series of instructions.

It is important to emphasise that if an instruction is wrong, it can easily be changed and that making mistakes is inevitable; it is part of learning how to teach the toy what to do.

Activity 2: Demonstration of simple turtle graphics program

Introduce the program, explaining how the screen shows an aerial view of the turtle (just like the programmable toy) and how the controls work. Ask if they can work out how to control the turtle's movement to guide it around the screen. Depending on the program used and the current classroom topic, the screen could be a map of a town, a map of islands, a crazy golf pitch, a map of space, etc. Once it is clear the children understand what is required, explain the next activity. The level of complexity can be adjusted dependent on the extent to which the children seem to grasp the basic features of the program.

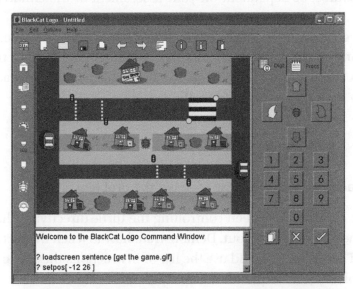

BlackCat Logo *with 'Get the Game' screen from the CD-Rom*

With most programs the controls refer to the turtle's orientation – that is, the turtle's left and forward, etc. However, some of the simpler control panels use the screen's orientation (up, left, right, down). The children will find it is possible to cheat by driving through houses etc., but you should emphasise that this will not help them learn.

Activity 3: Problem-solving scenarios

In pairs, the children work through one or more of the program's built-in scenes. You could give them free choice of the scene but define the level of complexity of the control panel used to guide the turtle. For example, *BlackCat Logo* program provides four built-in and one customisable level of control:

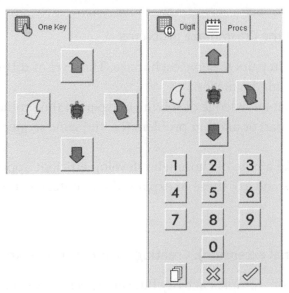

BlackCat Logo *control panels*

The levels of demand could be differentiated to reflect your knowledge of the children's capabilities or you could let the children select their own level of challenge. This will, of course, be dependent on your assessment of the children's levels of independence.

At this stage the turtle is under direct control – it is not remembering a list of instructions and so the children should make a note of their instructions so they will be able to 'teach' the turtle how to solve the problem in the next activity. They could develop their own shorthand methods – e.g.

$$\uparrow 3 \quad \rightarrow 2 \quad \nearrow \quad \uparrow 4$$

Activity 4: Teaching the turtle how to solve the problem

Until now the children have been controlling the turtle directly: as they give each instruction, the turtle carries it out. Demonstrate how to teach the turtle a series of instructions. The children should use the list of instructions they gave the turtle to solve the problem(s) in the previous activity.

Suggest that the children test their 'program' after entering two or three instructions in case they have made a mistake. You should be prepared to set more demanding challenges for those children who complete the activity with relative ease.

Activity 5: Demonstration of *Crash* program

Load the *Crash* program from the CD-ROM accompanying this book and, with the children gathered round, ask if anyone can work out how to guide the turtle through the first of the *Crash* mazes. Once the first screen has been solved, reload the program and see if anyone can solve the first maze in fewer moves.

The children should begin to notice that information used to solve one part of the maze will help them solve other parts (i.e. the distances travelled are the same for some parts of the maze).

Activity 6: Solving the *Crash* maze problems

The children work in pairs to solve each maze. The level of difficulty increases as they progress through the mazes.

Remind them that knowledge of solving one part of the problem can be used to help solve another part or another problem (i.e. the same sequence of instructions is likely to recur).

You should make a note of those who develop efficient approaches to solving the maze problems and ask these to explain their strategies in the end of session plenary.

Activity 6a (optional extension): Creating their own mazes to solve

This is a slightly more complex activity but can be very rewarding, particularly for those children needing more challenge (or those who are older than the target group). The first maze screen (i.e. *square1*) is loaded into a paint program and, provided the locations of starting and end points are not changed, and the same colours are used for the road (i.e. white) and the barriers (i.e. red), the children can create their own maze for others to solve.

Once the maze has been redrawn it is saved in the screens folder for the *SuperLOGO* demonstration program. When the program is started, the new maze is loaded for them to solve (see *What do I need to know?*, below).

What should the children know already?

How to control a programmable toy (see Project 1)

Most children will have used a programmable toy but some may need reminding about how it works. They also need to be reminded about how the toy follows a set of instructions in sequence and how the list can be modified if there is a problem. They should also be reassured that making a mistake is part of the learning process.

What resources will I need?

A programmable toy

Although not essential, the toy is useful to remind the children about how it is controlled and to help establish the relationship between the toy and the aerial view and control of the screen turtle. Any programmable toy can be used such as *Roamer*, *Bee-Bot*, *Pip* or *Pixie* (see Project 1 for more information).

A simple turtle graphics program

Unlike most full versions of LOGO, the simple introductory programs include a series of simplified controls graded for increasing levels of difficulty and usually a set of screens which the children can explore. Suitable programs include: *2Go* (2Simple), *Terry the Turtle* (Kudlian), *Textease Turtle* (Softease), *Roamer World* (Logotron), *Granada Logo* (Granada Learning), *BlackCat Logo* or BlackCat *Turtle 2*.

LOGO

The activities and the LOGO files on the CD-ROM accompanying this book can be used with any version of LOGO.

Crash

This *SuperLOGO*-based program (Logotron) is provided on the book's CD-ROM as a self-running file which can be installed on any Windows computer.

What do I need to know?

How to access and install online web-based activities

An evaluation version of *Terry the Turtle* can be downloaded from http://www.kudlian.net/products/terry/ and is also provided on the CD-ROM which accompanies this book.

To access these websites type the above web address into the address pane on your web browser (e.g. Internet Explorer, Firefox, etc.) and then press the **ENTER** key on your keyboard.

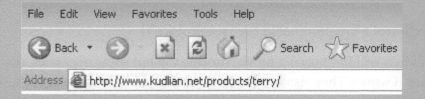

To install the demonstration program, click on the **Download** link and when presented with the install screen click on **Run.**

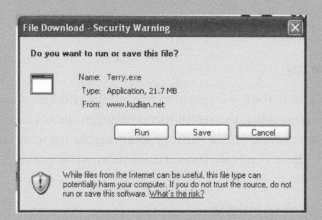

This will install the demonstration program on your computer which can then be accessed from **Start > Programs** in the usual way.

NOTE: As this is the demonstration version, it will run for only five minutes before shutting down. You will need to purchase the software and receive a registration code to run the fully functioning version of the software.

How to use a simple LOGO turtle graphics program such as *Terry the Turtle*, *2Go*, *Softease Turtle* or *Roamer World*

Once loaded, these programs are relatively straightforward to use.

Changing levels on *Terry the Turtle*
The initial level of control is decided through the selection of the various screens. These can be changed at any point by holding down the **Ctrl** key and typing **1, 2, 3** or **4.**

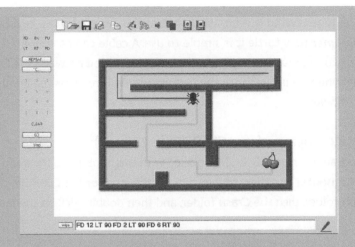

FD 12 LT 90 FD 2 LT 90 FD 6 RT 90

Level 4 requires the children to enter a series of instructions and press the **Go** button to see them carried out.

Changing levels on 2Simple *2Go*

The four levels of control and turning on the programming flow-chart feature is achieved by using the *Teacher controls*. Once the program has loaded hold down the **Ctrl** and **SHIFT** keys and then type the letter **O**. Clicking on the **General** tab allows you to choose the type of control panel shown to the children and also enables you to decide whether the children can program the instructions into a flow chart (see Activity 3).

Changing levels on *Softease Turtle*

The control panel or 'keypad' is shown by clicking on the **Keypad** button in the top left corner of the Softease window. The various levels of control panel can be chosen by clicking on the yellow, green or blue buttons at the top of the keypad window.

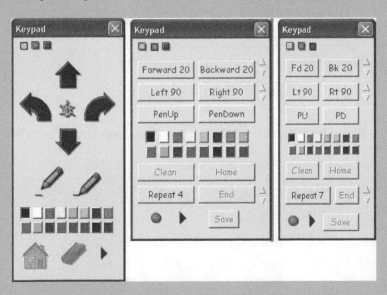

Note that the green and blue levels allow the children to 'record' a series of moves or instructions. Instructions can also be entered by creating a procedure (under the **Turtle** menu).

There is only one level to *Roamer World* but as this resembles the keyboard on the back of the *Roamer* floor turtle it is simple to use. A cable can be connected from the computer to a *Roamer* turtle and instructions can be passed between the *Roamer* and the computer which would enhance the first activity. This requires the purchase of the special cable, however.

How to install programs from CD-ROM (e.g. *Crash*)

The *Crash* program is on the CD-ROM provided with this book. From the desktop, double click on **My Computer** and then double click on the icon for the CD-ROM drive. Open the **Project 04** folder, then the *Crash* folder, and then double click on the **Install** file.

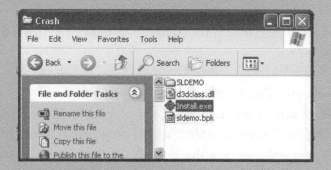

Once the program has installed it can be loaded in the usual way from the **Start >** **Programs** menu.

Click on the **Demo Projects** button, then click on the **Crash** button and click **OK**.

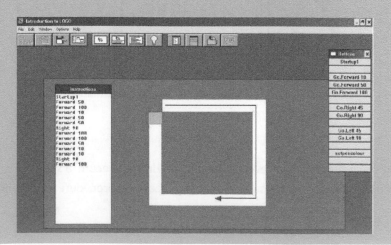

(Optional) How to create additional mazes for a turtle graphics program using a *Paint* program

You will need to be confident with working with files and images for this part. Open the **Screens** folder which, once *Crash* has been installed, should be in **C drive >** **SLDEMO > SCREENS**.

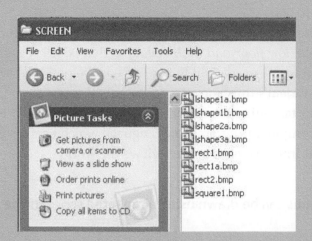

Open one of the bitmap images (e.g. *square1*) in any paint program (e.g. *Paint* in **Start > Programs > Accessories**). It might be advisable to save the original image with a different name (e.g. *square2*) before changing it.

Redraw the maze using exactly the same colours (i.e. red for the barriers and white for the roads) and make sure the position of the start and finish points is not changed. Then save the new maze with the same name as the original (i.e. *square1*). When you next run the program the new maze will appear in place of the original.

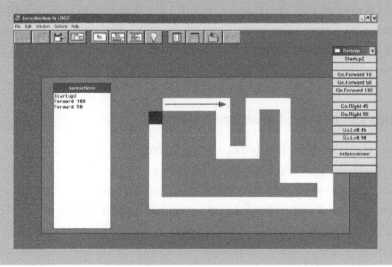

What will the children learn?

About body-centred geometry (i.e. left, right, forward, back)

The foundations of LOGO turtle graphics programming

This project is designed to bridge the gap between giving instructions to programmable toys and LOGO turtle programming. LOGO programming uses body-centred geometry as its basis, the learning principle being that children will identify with the turtle and through it be able to explore and develop their understanding of two-dimensional space and shapes. By visualising the drawing of a 2D shape from the turtle's perspective, the children gain deep insights into the properties of such shapes (see Seymour Papert, *Mindstorms: Children, Computers and Powerful Ideas*, Harvester, 1980, p. 55).

A major feature of LOGO programming is what Papert, the founder of LOGO, considers to be its '*low floor and high ceiling*' – that is, it is easy enough for the youngest, most inexperienced child to use but has the potential to challenge even a post graduate student's thinking.

The most important thing to remember about LOGO is that it is not a drawing program; it is a computer programming language which enables you, the teacher, to gain an insight into how each learner views the world.

That regular shapes can be drawn using a repeated series of the same instructions

And hence, the properties of some familiar regular shapes

Although most children can correctly identify and categorise shapes, differentiating, for example, a square from a rectangle, many are unable to define the properties of some very common shapes. For example, if you ask seven-year-olds what they can tell you about the sides of a square, many will not be able to tell you with confidence that all four are the same length – or if you tell a child that the length of one side of a square is 8 cm, he/she will not be able to use this information to tell you the length of any of the other sides of a square. The activities in this project are designed to help children appreciate some of these relationships – but they will need to be emphasised in end-of-session plenaries.

That problems can be solved in different ways

Although we are attempting to encourage children to look for the most elegant solutions to the maze problems, the fact that they can be solved in different ways should also be mentioned – and also that each solution has its merits, not least that it enables you, as the children's teacher, to gain an insight into how they perceive a shape and/or a situation. Some children are holistic thinkers and will see the maze in its entirety and hence will spot patterns in the solution of a problem; other children are serialist thinkers and work from one problem to the next, seeing each sub-problem as a separate entity. It may not be possible to change a serialist thinker into a holist or vice versa.

That LOGO procedures comprise a series of instructions which can be amended

Encouraging children to make a record of their attempts to solve the problems can help them develop a more systematic approach. If, for example, they correctly solve a maze problem apart from the penultimate turn, a record of the moves to that point will enable them to re-enter these instructions and change only the final one to solve the problem. This notion of 'debugging' not only their list of instructions but also their thinking is one of the key principles underpinning the constructivist approach to learning through LOGO.

Challenging the more able and supporting the less able: modifying the project for older and younger pupils

Structuring the progression of tasks to match the capabilities of the children

Observation of the children's approaches to the solution of the problems should enable you to gauge whether some can be fast-tracked to more demanding problems while others need to spend more time tackling those which are easier to solve. You may find that some children, who normally find other aspects of numeracy difficult, are more adept at these spatially oriented tasks – particularly some boys who, it has been found, develop spatial awareness more readily than some girls. The extension activity can be used to raise the level of challenge through the creation of more complex mazes (e.g. by requiring turns other than 90°).

Differentiating the level of support provided

This can be achieved through the use of well-briefed adults or could be managed through peer mentoring, provided the dynamics of the relationships within the class or between older and younger children are supportive of such an approach.

Why teach this?

LOGO programming not only enables the children to develop a systematic approach to solving problems by allowing them to see immediately the effect of their decisions; it also helps you as their teacher to gain an insight into how children's minds function. Some children will naturally be systematic and orderly in their approach, while others will be haphazard and slapdash. The aim should be to steer those who are less well organised into a more orderly approach – to help them recognise how the information they have gathered through their mistakes or the solution of previous problems can be applied to help solve similar problems.

In addition, they are gaining an insight into how computers function – i.e. they blindly carry out the instructions they have been given.

In the existing QCA scheme of work for ICT there is a conceptual and sequential gap between Unit 2D: *Routes: controlling a floor turtle* and Unit 4E: *Modelling effects on screen*. This project enables the children to build upon understanding developed in Unit 2D (or Project 1) in preparation for or in place of Unit 4E.

Making and testing predictions about comparative distances and angular measures helps reinforce the children's understanding of numbers and the number system, as well as helping them develop orderly and systematic approaches to solving problems. The activities in this project lay the foundations for the creation of 2D shapes and the accurate measurement of angles (as the measurement of turn). The use of LOGO reinforces the connection between measurement, shape and numerals.

These activities are particularly valuable in helping children appreciate the relationship between angular measurement and the amount of turn. Inevitably, children will direct the turtle to turn in the wrong direction and hence will have to correct this by turning two right angles to face the opposite direction. They will also come to recognise that rectangles have two opposite sides the same length and squares have all four sides the same length.

See also *Arts* Project 8 (*LOGO animation*), *Arts* Project 9 (*Controlling external devices*) and *Humanities* Project 2 (*Map making using GIS*) for related activities.

Project Fact Card: Project 5: Symmetry and tessellation

Who is it for?

- 7- to 9-year-olds (NC Levels 2–4)

What will the children do?

- The children will initially investigate the symmetry of plane shapes through the use of a practice program and/or website. They will then explore reflective and rotational symmetry through the use of a drawing program. Finally they will investigate tessellation using and creating shapes which they can reflect (flip) and rotate

What do I need to know?

- How to locate and download web-based activities for classroom use
- How to use basic computer drawing tools – particularly the flip and rotate features
- How to create tessellating shapes with a paint program
- How and why shapes tessellate and the connection with symmetry

What should the children know already?

- The names of the most common plane shapes (e.g. square, rectangle, triangle, circle)
- That all three-sided shapes are called triangles, four-sided shapes are quadrilaterals, five-sided shapes are pentagons, etc.
- That regular polygons have all sides (and angles) the same
- How to use a mouse to point, select and drag and drop

What resources will I need?

- A symmetry practice program or website
- A vector drawing program (or drawing tools)
- A paint program such as BlackCat *Fresco* or BlackCat *Painter 2*, or Sherston's *Tessellation Exploration* program

What will the children learn?

- Which shapes tessellate and how
- About the reflective and rotational symmetry of plane shapes
- How to use a drawing program (or tools) to explore symmetry and tessellation
- How to design tessellating shapes using a paint program or a dedicated tessellation or tiling program

How to challenge the more able

- Investigate more complex tessellations in a systematic way (e.g. semi-regular tessellation)
- Allow free exploration of the tessellation programs/websites

How to support the less able

- Simplify the tessellations they explore
- Structure the activities more carefully to provide step-by-step scaffolding of tasks
- Provide more adult (or peer) support

Why teach this?

- It introduces the children to ICT NC KS1 PoS statements 1a, 2a, 3a; KS2 PoS statements 1c, 2c, 3a
- It replaces QCA ICT Scheme of Work Unit 3D and complements Unit 4B
- It addresses Mathematics NC KS1 PoS statements Ma3, 1a, 1c–f, 2a–d, 3a, 3b; KS2 PoS statements Ma3, 1c–e, 1g, 1h, 2a–c, 3a, 3b
- It reinforces NNS Y3 Au/Su/Sp Units 4–6; Y4 Au/Su/Sp Units 4–6

Symmetry and tessellation

What will the children do?

Activity 1: *RoboPacker*

Load up one of the *RoboPacker* games appropriate for the age group and capabilities of the children you are teaching (see *What resources will I need?*, below). Ask them if they can work out how to play the game. Make sure all the controls are demonstrated, particularly rotate and reflect. Emphasise that there is sometimes more than one correct way of packing the robots.

Send them to work in pairs. The program automatically moves them on to the next level but some of the children might reach the topmost level for the grade selected. If this is the case, you could load the game for the next grade (or even Grade 6).

In the plenary at the end of the lesson ask the children to comment on the game and what they noticed about the shapes of the robots and whether there was more than one way of packing the robots into the cases. For example:

Ask the children if they have noticed what shapes the robots are (they are triangular, quadrilateral or hexagonal) and then ask if they can decide why that might be. This leads into the investigation in the next activity.

Activity 2: Investigating tessellations with regular polygons

You need to download the *Polygons Around a Point* program for this activity (see *What resources will I need?*, below).

Recap on the *RoboPacker* activity, reminding the children about the two questions posed about the shapes being packed. Demonstrate how to use the *Polygons Around a Point* program and how to fill in the recording sheet. (NOTE: In case the sheet provided with the program is too complex, a simplified version is available on the CD-ROM accompanying this book.)

The children investigate the polygons in pairs, finding out which will or will not tessellate around a point, and record their findings. They could use the accepted shorthand method (e.g. *triangle, triangle, square, hexagon* = 3.3.4.6 or *hexagon, hexagon, hexagon* = 6.6.6).

In the end of lesson plenary, the children share their findings and you draw out any generalised conclusions. For example:

◉ The only polygons which tessellate with themselves (i.e. regular tessellations) are the triangle, square and hexagon.

◉ The semi-regular tessellations are: (3.3.3.4.4), (3.3.4.3.4), (3.4.6.4), (3.6.3.6), (4.8.8), (3.3.3.3.6), (3.4.4.6), (5.5.10). Be aware that some children may appear to find others because they have not quite followed the convention (e.g. 4.8.4 is the same as 4.4.8).

NOTE: There are three other semi-regular tessellations, but these involve dodecagons which do not feature in the program.

Optional extension/homework activity

The children could colour in semi-regular tessellations using the templates provided on the *Totally Tessellated* website:

http://library.thinkquest.org/16661/ templates/index.html

Activity 3: Investigating tessellations with a drawing program

Prepare a drawing program (or the drawing tools in *MS Word*). Remind the children what is meant by a tessellation – i.e. shapes fill an area with no gaps. This could be done by accessing one of the tessellation websites, one of the examples from this book's CD-ROM, or with a 'one I made earlier' example. Give them a challenge to investigate: *'Is it true that any triangle, quadrilateral or hexagon will tessellate with itself to fill a space?'* Demonstrate how to use the drawing tools, particularly how to copy and paste a shape, how to fill a shape with colour and how to rotate and/or flip a shape.

In pairs, the children investigate tessellations with one shape, printing out and saving their work. If possible, they could save their findings to a shared folder on the network or a memory stick so they will be able to share what they have discovered in the plenary.

At some point in the session, you could demonstrate how any triangle or quadrilateral can be tessellated by copying and pasting the original shape, then flipping horizontally and vertically and then joining the two shapes together by grouping. The resultant merged shape will tessellate quite easily:

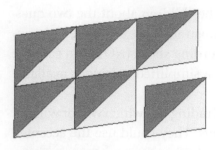

A double-flipped triangle combined with itself produces a quadrilateral which tessellates easily

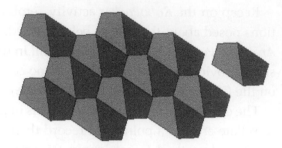

A double-flipped quadrilateral combined with itself produces a hexagon

It might be useful for the children to be shown how to group shapes together (see *What do I need to know?*, below).

Activity 4: Rotational and reflective symmetry

Show the children the presentation on line (reflective) symmetry on:

http://www.linkslearning.org/Kids/1_Math/2_Illustrated_Lessons/4_Line_Symmetry/index.html

After a brief discussion in which the key points are drawn out, show the children how to access and use the *Symmetry Game* which they then work in pairs to complete:

http://www.innovationslearning.co.uk/subjects/maths/activities/year3/symmetry/shape_game.asp

You could suggest that if the children do not achieve a perfect score the first time, they work their way through the quiz again.

Once all have completed the quiz, demonstrate how to rotate a shape using drawing tools and explain the rotational symmetry e-worksheet (provided on the CD-ROM for this book) which they then complete.

In the final plenary, draw out any generalised conclusions from the children's findings (for example, they may predict that for any regular shape the order of symmetry is the same as the number of sides). Note also that:

⊙ 'line symmetry' is the same as 'reflective symmetry';

⊙ 'flipping' a shape is the same as reflecting it.

Activity 5: Creative tessellations

Show the children examples of M. C. Escher's tessellated artwork, e.g. using the M. C. Escher website:

http://www.mcescher.com/Gallery/gallery-symmetry.htm

Ask if they can identify the polygons on which the designs are based.

In pairs, let the children explore the tessellations on the *Tessellation Town* website:

http://www.mathcats.com/explore/tessellationtown.html

Discuss what the children have noticed about how simple tessellating shapes can be modified to produce more complex ones. Show the children how a triangle or quadrilateral can be manipulated to create tessellations, using the Shodor Educational Foundation *Tessellate!* webpage:

http://www.shodor.org/interactivate/activities/tessellate/index.html

In pairs, let the children create their own tessellations using the *Tessellate!* website and then, in a plenary, discuss what the children have discovered. For example, many complex tessellations are based on the regular and semi-regular tessellations the children have been investigating.

Activity 6 (optional extension): Creative tessellations 2 (with a paint or tessellation program)

This activity could be incorporated with or replace Activity 5. The advantage of this activity is that the children will be able to save and print out their designs and, if a paint program is used, it will develop the children's ICT capabilities more extensively. It also contributes to the development of ICT capability through the use of a paint program. Paint programs differ from drawing programs – shapes on a drawing program remain as separate 'objects' whereas shapes become part of the background in a paint program.

Recap on what the children have discovered so far about tessellation and symmetry. Show the children some of the artwork created by M. C. Escher or similar. Demonstrate how to create tessellations using a paint program (see *What do I need to know?*, below) or a dedicated tessellation program.

In pairs, the children produce their own tessellations using the software and then, in the plenary, discuss what they have found out about shapes and the software they used to produce their designs.

What should the children know already?

The names of the most common plane shapes (e.g. square, rectangle, triangle, circle)

Although it is not essential the children know the names of basic shapes, it will enable them to communicate their findings if they are already familiar with them.

That all three-sided shapes are called triangles, four-sided shapes are quadrilaterals, five-sided shapes are pentagons, etc.

This can be taught or reinforced through the project but it would help avoid confusion if the names of polygons are clarified in some way before or at the start of the project.

That regular polygons have all sides (and angles) the same

You might find that some children assume, for example, that 'hexagon' applies only to a regular hexagon. They may need to be reassured that any six-sided shape is a hexagon – even these:

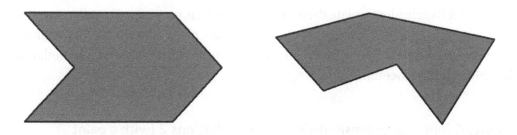

How to use a mouse to point, select and drag and drop

The activities in this project will help children lacking in experience to consolidate their basic skills but it should not be assumed that all children are equally adept at co-ordinating hand and eye when using a mouse.

What resources will I need?

A symmetry practice program or website

RoboPacker

Houghton Mifflin provides a range of educational activities on their website; their *RoboPacker* game is highly appropriate as an introductory activity for this project. There are six versions of the game and six levels of difficulty within each game. The Grade 1 (simplest) game can be accessed by typing in this address:

http://www.eduplace.com/kids/mw/swfs/robopacker_grade1_t.html

The Grade 6 game can be accessed through this address:

http://www.eduplace.com/kids/mw/swfs/robopacker_grade6_t.html

The other games can be accessed by changing the grade number in the address.

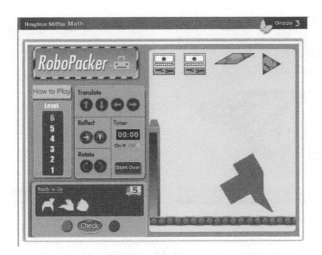

The game requires the children to pack the various shaped robots into oddly shaped packing cases on the conveyor at the base of the screen. This may require some robots to be rotated or reflected to fit into the case.

NOTE: Once loaded the program will operate successfully even if the computer is disconnected from the internet. You might find it useful to load the program on all the children's computers before the lesson begins in case there is a problem with the internet.

Polygons Around a Point

The *Polygons Around a Point* program can be downloaded from:

http://www.harcourtschool.com/clab/download_7.html

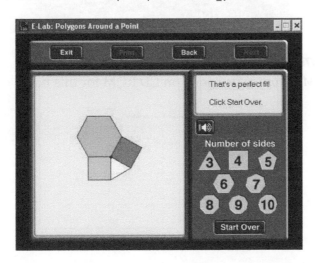

You will need the E-Lab application program and the *Polygons Around a Point* file (numbered Chapter 11) – both available on the same web page.

Line Symmetry

An online interactive presentation for children on line (reflective) symmetry is provided by Links Learning – an American educational organisation:

http://www.linkslearning.org/Kids/1_Math/2_Illustrated_Lessons/
4_Line_Symmetry/index.html

 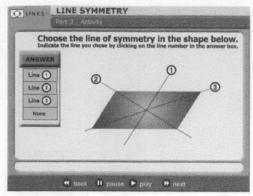

Symmetry Game

The (reflective) *Symmetry Game* is available online at:

http://www.innovationslearning.co.uk/subjects/maths/activities/year3/
symmetry/shape_game.asp

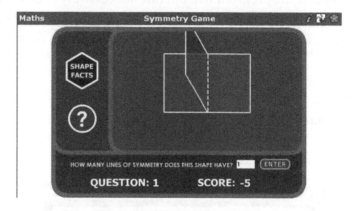

Rotational Symmetry e-worksheet

An example for *MS Word* is provided on CD-ROM for this book.

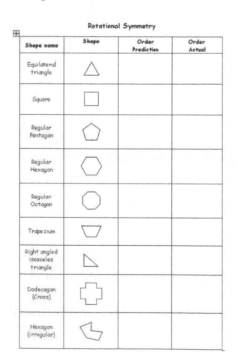

Rotational Symmetry

Shape name	Shape	Order Prediction	Order Actual
Equilateral triangle			
Square			
Regular Pentagon			
Regular Hexagon			
Regular Octagon			
Trapezium			
Right angled isosceles triangle			
Dodecagon (Cross)			
Hexagon (irregular)			

Alternatively, you could produce a similar version using your preferred drawing program.

Tessellate!

The Shodor Education Foundation website provides a very good online activity for manipulating shapes to create tessellations:

http://www.shodor.org/interactivate/activities/tessellate/index.html

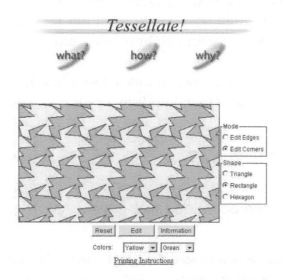

Tessellation Town

Tessellation Town on the Math Cats website includes a set of simple tessellation tasks which should stimulate the children's thinking and imaginations. The *Tessellation Town* activity can be accessed via:

http://www.mathcats.com/explore/tessellationtown.html

A vector drawing program (or drawing tools)

Vector drawing programs differ from painting programs in that the objects which are drawn on screen remain as separate objects which can be readily moved, resized, reflected or rotated. In a painting program once a shape has been placed on the screen it remains fixed to the screen. Drawing programs which would be suitable for this project include *aspexDraw* (Aspex), *Granada Draw* (Granada Learning), *Oak Draw* (Dial), *Access Maths 4* (ACE), *Textease Draw CT* (Softease).

Alternatively, the drawing tools in *MS Word* or *Textease* can be used.

A paint program or Sherston's *Tessellation Exploration* program

A basic paint program will suffice, including the *Paint* program which is provided with your computer (**Start > Programs > Accessories > Paint**). Alternatively, Sherston Software's *Tessellation Exploration* program is relatively inexpensive and is designed specifically for creating free-form tessellations such as those devised by the artist M. C. Escher (see http://www.mcescher.com/Gallery/gallery-symmetry.htm).

Alternatively a tessellation program called *Tess*, which is similar to Sherston's program, can be downloaded from http://www.peda.com/tess/

What do I need to know?

How to locate and download web-based activities for classroom use

Detailed information on how to access websites is shown in Project 4 and guidance for downloading web-based activities is included in the *What do I need to know?* section of Project 3.

How to use basic computer drawing tools – particularly the flip and rotate features

The guidance provided here applies to the drawing tools provided with *MS Word*, but most vector drawing tools work in the same sort of way.

⊙ To show the drawing tools in *Word*, click on the **View** menu and then select **Toolbars > Drawing**.

- This will display the **Drawing** toolbar which can be positioned anywhere on the screen. The most usual place for it to be positioned is at the base of the screen. To do this, drag the toolbar to the base of the screen.
- To draw a shape on screen, click on **AutoShapes** on the **Drawing** toolbar and select the type of shape you require – in this example, a *hexagon* has been selected:

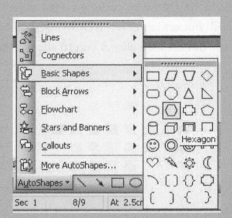

- Move the mouse pointer to any point on the screen and drag out the shape to the size you require. A hexagon should appear. (NOTE: If you require a regular hexagon as opposed to the squashed or stretched version, hold down the **SHIFT** key as you drag out the shape.)
- Drag on the white 'handles' on the corner of edges of the shape to resize it.
- To 'flip' (reflect) the shape, click on the shape to highlight it, then from the **Draw** menu on the **Drawing** toolbar, select **Rotate or Flip**.

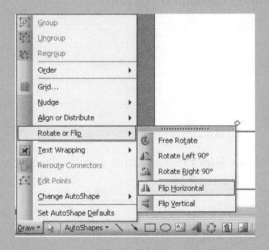

- You can also rotate the shape using this option. Alternatively, a shape can be rotated by highlighting it and dragging the green handle round in an arc.

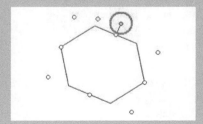

⊙ Several objects (shapes) can be grouped together to form a single object by highlighting all the objects to be grouped (click on an object and then hold down the **Ctrl** key and click on the other objects required). Then, from the **Draw** menu on the **Drawing** toolbar, select **Group** (or right-click and select **Grouping >** **Group**).

⊙ An object can be filled with colour by clicking on the object to highlight it and then clicking on the **Fill** bucket icon on the **Drawing** toolbar. Clicking on the down arrow beside the bucket icon allows you to choose the fill colour:

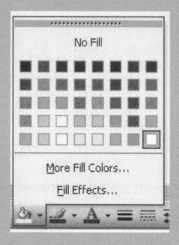

How to create tessellating shapes with a paint program

⊙ Open the *Paint* program
⊙ Draw a rectangle in the corner of the screen.

⊙ Click on the **Free-form Select** tool and then click the *Transparent Background* button.

⊙ With the **Freeform Select** tool, carefully cut out a shape, starting from one edge and returning to the same edge.

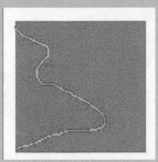

⊙ Drag this selected area to the opposite side of the shape.

⊙ Repeat this process for the other edge.

⊙ Copy the original shape and paste it somewhere else on the screen.

⊙ Fill this shape with another colour.

⊙ Select this shape and drag it to tessellate with the original shape.

⊙ Select the new combined shape then copy and paste this shape.
⊙ Drag the pasted shape to tessellate with the original shape.

⊙ Repeat the paste and tessellate process until you have filled the entire screen.

More detailed guidance for the *Paint* activity is provided at:

http://www.wsd1.org/bitsbytes/9798/bboct97/default.htm#STORY4

> ## How and why shapes tessellate and the connection with symmetry
>
> A short summary of information on tessellations sufficient for this project is provided on this website:
>
> http://mathforum.org/sum95/suzanne/whattess.html
>
> For further background information refer to these websites:
>
> - ⊙ *Totally Tessellated*: background information on symmetry and tessellation – http://library.thinkquest.org/16661/index.html
> - ⊙ More generalised background information on geometry – http://homepage.mac.com/efithian/geometry.html

What will the children learn?

Which shapes tessellate and how

As can be seen from the information on the websites listed above, there is considerably more about tessellation which could be studied. You might, for instance, lead on to a more in-depth study linking with religious education (Islamic patterns) and art. This project could be followed up as the children progress through the school to investigate more systematically and mathematically the relationships between the shapes and the angles where the shapes meet (i.e. they must all add up to 360°), which is why only triangles, squares and regular hexagons tessellate (60°, 90° and 120° are factors of 360°).

About the reflective and rotational symmetry of plane shapes

Often symmetry is tackled as a topic with no particular reason other than it is part of the curriculum. Incorporating symmetry with tessellation provides a more interesting context for finding out about symmetry. Shapes which remain unchanged when they are flipped or rotated in a drawing program are those with reflective and/or rotational symmetry.

How to use a drawing program (or tools) to explore symmetry and tessellation

By using both painting and drawing programs in this project the children will come to appreciate the difference between the two and the advantages and disadvantages of each. For mathematical investigations, drawing programs offer more scope. Through this project and related projects in the other subjects, children will develop an appreciation of ICT tools and which are most appropriate for particular tasks.

How to design tessellating shapes using a paint program or a dedicated tessellation or tiling program

The brief investigation into Escher's designs in the final activity provide a taster which some of the children may decide to follow up in their own time. Pursuing some of the links to related websites can prove fascinating, showing quite distinctly the relationship between mathematics and art.

Challenging the more able and supporting the less able: modifying the project for older and younger pupils

Adjusting the level of challenge or support provided by the task

Children capable of more demanding work could explore some of the activities on the websites listed above. For example, two successful solutions to Activity 2 (*Polygons Around a Point*) do not tessellate to fill an area. The children could be asked to investigate these tessellations. Similarly, this project looks only at reflective and rotational symmetry, but some tessellations involve translational and glide reflective symmetry.

Some children, regardless of their other abilities, may become fascinated by tiling and tessellating. Two more online resources which they might like to explore include:

⊙ Download free *QuiltMaker* software – http://www.quiltmakersoftware.com/

⊙ Online pattern blocks activity – http://ejad.best.vwh.net/java/patterns/patterns_j.shtml

As with all projects, the level of challenge and support (adult or peer) can be adjusted to meet the needs of the children.

Why teach this?

The investigations and interactive tasks make use of ICT to provide information or generate data and most importantly to develop the children's ideas through the manipulation of information in simulations. The *RoboPacker* activity is a problem-solving simulation in an imaginary context. The children are required to test out their ideas and check the results of their decisions through feedback provided by the computer. The *Polygons Around a Point* program is another simulation – it enables the children to investigate the effects of combining different polygons accurately and gain feedback on the effects. This enables them to focus more on the gathering and recording of data rather than the mechanistic process of having to draw around cut-outs of shapes. The *Symmetry Game* is another form of simulation in that if the children make mistakes, the reflection is modelled on screen for them. The paint and draw activities automate the processes of experimentation which could be carried out more laboriously with pencil, templates, card and scissors. These activities provide the children with an environment which is less prone to error or confounding variables, again enabling them to focus on the key learning objectives. All the while, the children are learning how ICT tools can assist them with other tasks.

QCA ICT Unit 3D: *Exploring simulations* is somewhat vaguely worded, leaving you to work out for yourself the simulation(s) which might be appropriate and the contexts in which they might be used. This project makes use of several types of simulation in purposeful contexts, helping children appreciate that simulations are used in the real world to enable people to test out designs and ideas before committing themselves to costly or dangerous projects.

The difference between a pattern and a tessellation is that the shapes in tessellations fit together without spaces, whereas a pattern comprises a series of repeated designs in which the spaces are uniform. ICT Unit 4B: *Developing images using repeating patterns* focuses on pattern making and refers to Islamic designs and symmetry without specifically mentioning tessellation. Some of the ideas from Unit 4B could be combined with this project or vice versa.

This project provides a platform for addressing many of the statements in Ma3 Shape, Space and Measures. Embedding the activities into purposeful investigations with interesting open-ended outcomes stimulates interest and helps children appreciate the relationship between numbers and shape. Because shape is a visual medium, children should be given ample opportunities to explore patterns and relationships involving numbers and shapes to help them visualise how numbers work.

This project provides children with an opportunity to investigate and make reasoned predictions and conclusions about the way shapes work. They may already be familiar with shapes and know that some fit together neatly while others do not. By investigating this implicit knowledge in a more systematic way, they not only find out more about these shapes but also are learning about the importance of being organised and recording their findings to communicate with others. The use of ICT resources assists in this process by taking away some of the laborious aspects of the tasks to enable the children to focus on the results and their significance. You will notice that all the investigations are enquiry-led – that is, they are stimulated by a question or designed to verify a statement. Opportunities should be taken to ask the children if they have ideas for how the investigations should be conducted throughout the project.

See also *Arts* Project 2 (*Aboriginal art*) and *Arts* Project 4 (*Designing logos*) for related activities.

Who is it for?

- 7- to 9-year-olds (NC Levels 2–4)

What will the children do?

- After working as a class to investigate the relationship between gender and favourite colours, the children will work in pairs (or threes) to devise and carry out their own enquiry. After creating a data-gathering sheet on a word processor and gathering data to address their enquiry, they will create a database file, enter the data, analyse them to answer their question and then present their findings for others. Finally, they will evaluate what they have done and apply their knowledge to data gathering and analysis in the real world

What should the children know already?

- That numerical information can be presented as graphs on a computer
- How to incorporate images and text in a word-processed document

What do I need to know?

- About the data-handling cycle
- How to create a simple database
- How to search for records which meet particular criteria in a database
- How to present information as bar charts in a database
- How to save a chart/graph as an image
- How to insert an image in a word processor document

What resources will I need?

- A database program suitable for use in primary school (which includes the facility to analyse selected records with charts/graphs and can export graphs as images) such as *Granada Database* or BlackCat *Information Workshop 2000*
- A word processor which can import images such as BlackCat *Writer 2*

What will the children learn?

- How to create a simple database
- How to search for records which meet particular criteria in a database
- How to present information as bar charts in a database
- How to transfer a chart/graph as an image from a database to a word-processed document.

How to challenge the more able

- Encourage the children to pursue a more challenging enquiry
- Investigate a follow-on enquiry

How to support the less able

- Scaffold the activities more systematically
- Provide more support (adult and/or peer)

Why teach this?

- It covers ICT NC KS2 PoS statements 1a–c, 3a, 3b
- It complements or replaces QCA ICT Scheme of Work Units 3A and 3C and augments Units 4A and 4D
- It addresses Mathematics NC KS2 PoS statements Ma4, 1a–c, 1e, 2a–c, 2f
- It reinforces NNS Y3 Unit Au/Sp/Su 5; Y4 Unit Au/Sp/Su 5

Statistical investigations 1

What will the children do?

Activity 1: A whole-class investigation – do girls prefer warm colours and boys prefer cold colours?

Ask the children whether they have used a computer to produce graphs. Check what they have graphed and what advantages using a computer might have over doing them by hand. It is likely that in most cases, they will have produced graphs of favourite football teams, pop artistes, foods, fruits, toys, etc.

A pre-prepared database is used to save time and to focus on the key learning objectives. For example, the children's names and gender could have already been entered before the lesson.

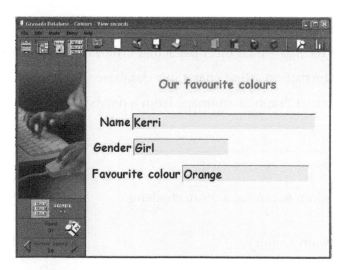

The children are asked for their favourite colour and this information is entered. The children could take turns to enter their own data or a scribe could be used.

Once the data have been entered, ask the children if they can suggest how they could find out which are the most and least popular colours. Hopefully, they will suggest graphing the colours for the whole class.

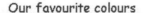

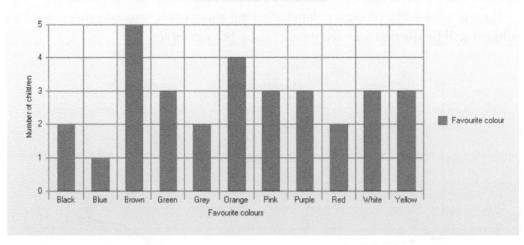

Our favourite colours

Ask the children if they know the difference between cold and warm colours. Pose the question, 'Do you think it is true that girls are more likely to select warm colours?' (and hence boys cold colours). Ask if they can suggest how they could use the database to find out.

Show them how to select the girls and then draw a graph of the girls' colours. Analyse the graph – have most of the girls chosen warm colours?

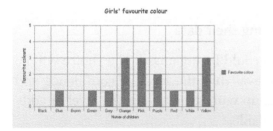

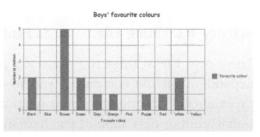

Repeat the process for the boys to see if the hypothesis is upheld.

Explain that in the next activity, the children will be working in pairs to pursue an enquiry similar to this one. They must start with an enquiry about children's favourite .... and then add an extra enquiry. For example: Favourite foods – do children who prefer chips watch more television? Favourite football teams – is there a link between favourite teams and favourite colour (e.g. do Liverpool fans prefer red)? Favourite books – do Harry Potter fans like travelling by train? Favourite pets – do girls prefer cats and boys prefer dogs? Favourite animals – do girls prefer soft and cuddly and boys prefer fierce and wild? Favourite snacks – do sweet eaters have more fillings in their teeth? etc.

Activity 2: Planning their investigation

Tell the children they must keep their hypotheses secret, as this might affect the answers the children give to their questions. They must then decide what data need to be gathered – for example, if investigating the link between favourite football team and colour, the children will need to gather those two pieces of information. They may decide also that they need the names of the respondents.

They now need to prepare a data-gathering sheet using a word processor. One column will be needed for each item of data. For example:

Name	Favourite team	Favourite colour

You could provide them with a template if you feel the children do not need to practise creating tables – it might even have the names of the children in the class already entered in the first column.

The children then gather their data. You could restrict the number of subjects they question to, say, 20.

Activity 3: Creating a database and entering their data

There are at least three ways this activity could be organised. The choice will be dependent on the equipment you have available, the level of confidence and experience of the children, your relationship with them and/or the level of additional adult or peer support you have available.

1. *Whole-class demonstration/pair work:* The process of creating a database is demonstrated to the whole class who then go to their computers and replicate the procedure.

2. *Step-by-step instruction:* You break the creation process down into a series of short sequences and demonstrate each step then let the children complete it before showing them the next step.

3. *Following a manual:* An illustrated worksheet is produced guiding the children through the process of creating their database.

You could combine the approaches by, for example, demonstrating the whole process (or part of it) and then getting them to work through a worksheet.

Once the database has been created the children work with a partner to enter their data. The advantage of working in pairs is that one can dictate the information while the other enters it.

In the plenary draw upon their experiences and reinforce the following:

⊙ A card is called a 'record', a chunk of information on a record is called a 'field', and the information entered into a field is called an 'entry'.

⊙ Once data have been entered into a database it is called a 'file'.

⊙ It is important that the data are entered accurately as this will affect the outcome.

Activity 4: Analysing the data and presenting the findings

The same organisational approach(es) could be followed as above, though if the children are already familiar with combining text and images in a word processor, the level of support may not need to be so great as with the previous activity.

The children analyse the data and produce the graphs needed to test their hypotheses. Dependent on the database/word processor combination used they will either have to copy or save their graphs and paste or insert them into their word processor report.

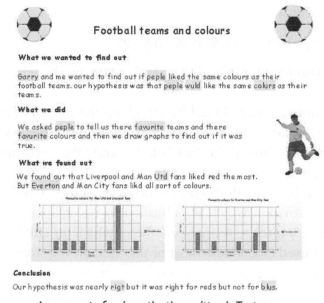

An account of an investigation written in Textease

If necessary, a writing frame or template document could be prepared with a word bank for those children needing additional support.

To complete the above enquiry, two searches needed to be combined: those liking Liverpool were combined with those liking Man Utd (searching for Liverpool OR Man Utd in the Team field).

Activity 5: Presentation of results and evaluation

Once all the children have completed their investigations and written their reports they could share their findings with the rest of the class. Because the investigations are different and the children have been involved in providing the data for the enquiries, they are likely to be very interested in the results.

Ask the children if they can suggest ways in which their enquiries could be pursued further. For example, the football teams/colours investigation could be followed up by seeing if people's favourite colours affect what they wear.

In conclusion, ask the children if they can work out how some of the following 'statistics' might have been discovered:

- Baby robins eat 14 feet of earthworms every day!

- Men are 6 times more likely to be struck by lightning than women!

- The Eiffel Tower in Paris weighs over 1000 elephants.

- 13% of Americans actually believe that some parts of the moon are made of cheese.

- The average chocolate bar has 8 insects' legs in it.

- During your lifetime, you'll eat about 60,000 pounds of food – that's the weight of about 6 elephants.

- There is enough graphite in the average pencil to draw a line 35 miles long – that's 45,000 words.

- Men get hiccups more often than women do.

- Rice is the main food for half of the people of the world.

NOTE: These 'statistics' were found by searching the internet (see, for example: http://www.excite.co.uk/directory/Recreation/Humor/Useless_Pages/Trivia). However, be warned: some of the 'facts' on this and other websites are not appropriate for children.

Finally explain that the information which the children have created is just as legitimate as information which is published in newspapers (or placed on the internet) and the approach to testing hypotheses which the children have just used is similar to that used to generate most of the 'statistics' featuring in newspapers and on TV (i.e. drawing generalised conclusions from a sample).

What should the children know already?

That numerical information can be presented as graphs on a computer

It is not essential that children have used computers to draw graphs but it is highly likely they will have done so at some point. It is advisable to not only check what they have already done but also to determine what they learned from this experience.

How to incorporate images and text in a word-processed document

The children ought to have word-processed documents containing images but may have used only built-in Clip Art or may have to be reminded about how to transfer images from one application to another. For example, where will the images be saved?

What resources will I need?

A database program suitable for use in primary school (which includes the facility to analyse selected records with charts/graphs and can export graphs as images)

A wide range of database programs is available for use in the primary classroom but some are restricted in the features they provide. For this project it is important that the database not only enables you to select particular records but also then allows you and the children to graph the records in the selection. The database should also enable you to save the graphs as separate image files.

The following databases are known to meet these criteria, but this is by no means an exhaustive list: *Granada Database* (Granada Learning), *Information Workshop 2000* (BlackCat), *Softease/Textease Database* (Softease), *Information Magic* (RM), *Junior Viewpoint* (Logotron). *Granada Database* is used as the exemplar in this project but all the databases listed work in similar ways.

A word processor which can import images

Most of the more recent educational word processors can import images, as will, of course, *MS Word*. *Textease* is used in this project, but most word processors will be able to achieve the same results.

What do I need to know?

About the data-handling cycle

Underpinning this project is the data-handling cycle:

The start of the process is the question or hypothesis; this gives a purpose to gathering the data. The analysis is the part played by the database. The interpretation of the data should lead back to the original question or in a spiral to the posing of another question or the formation of a new hypothesis.

How to create a simple database

The process described here uses the *Granada Database*, but all educational databases work in the same sort of way. *Textease Database*, for example, uses a Wizard which guides you and the children through the creation process. Example data files in various formats (*Textease Database*, *Granada Database*, text, csv, tsv) are provided on the CD-ROM that accompanies this book to enable you to experiment or to demonstrate completed files to the children.

1. *The database is given a name.*

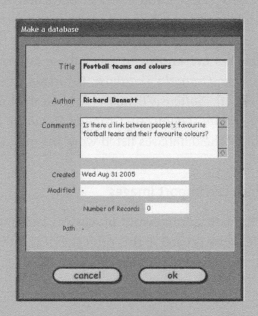

2. *The fields are created.*

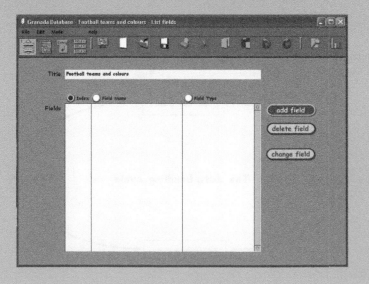

3. *The fields are defined (e.g. to contain text, numbers or a list of items).*

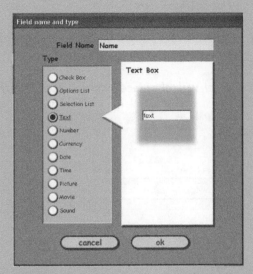

4. *The format for the screen is designed.*

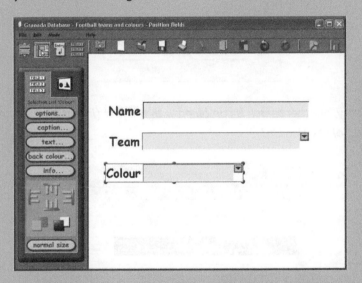

5. *The data are entered.*

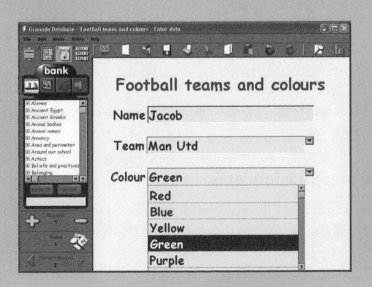

How to search for records which meet particular criteria in a database

The way searches are carried out varies slightly between databases but the process is largely the same.

1. *The field in which the search will take place is chosen and the search term is defined.*

2. *If necessary, another search is combined with the first (in this case we are searching for Liverpool OR Man Utd).*

Pressing the **Search** button will result in only the records containing the required information being shown.

How to present information as bar charts in a database

For this investigation, the graph will be for the selected records only. With some databases, you will also be asked to choose the type of graph needed at this stage in the process.

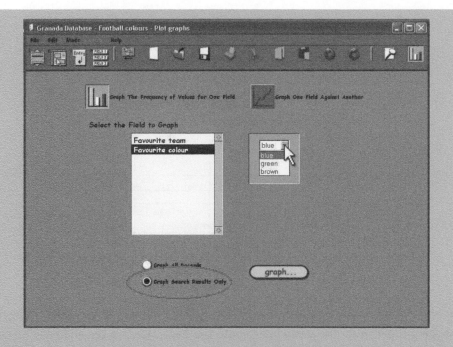

How to save a chart/graph as an image

Usually, clicking on the **File** menu will provide you with a **Save As ...** option. With some databases you need to click on a floppy disk icon to save the graph. In this example, the graph is being saved as a JPEG image as this takes up less memory and is an image format which can be inserted into most word processors. The image is being saved on the desktop so it can be easily located for inserting into the word processor.

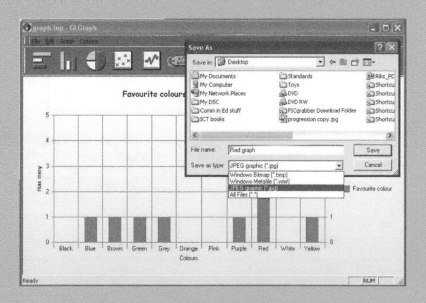

How to insert an image in a word processor document

Educational word processors vary in the process for inserting images. Some require the children to click on the **Insert** menu and then navigate to where the image has been saved.

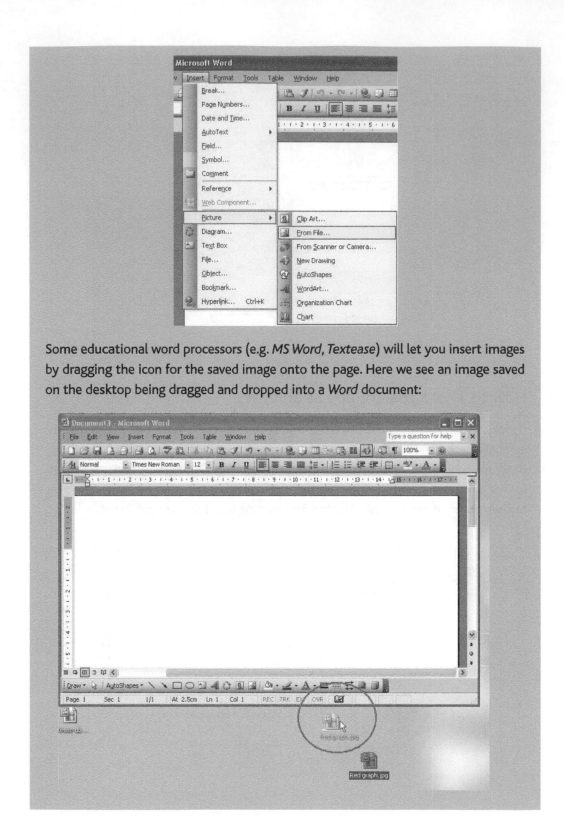

Some educational word processors (e.g. *MS Word*, *Textease*) will let you insert images by dragging the icon for the saved image onto the page. Here we see an image saved on the desktop being dragged and dropped into a *Word* document:

What will the children learn?

How to create a simple database

How to search for records which meet particular criteria in a database

How to present information as bar charts in a database

How to transfer a chart/graph as an image from a database to a word-processed document

These skills could be taught in isolation but this project is designed to embed them into purposeful activities. In this way the children will not only be acquiring the skills but also learning how and why they might be useful. This project can be followed up with Project 10, which builds on these skills and raises the level of challenge in terms of data handling.

Challenging the more able and supporting the less able: modifying the project for older and younger pupils

Varying the level of support provided

This can be achieved by structuring the tasks: those with more experience will be able to work more independently after an initial briefing, whereas those with less confidence will require a more step-by-step approach. As indicated in Activity 3, a manual could be prepared taking the children through the process of creating a database using the sort of screen shots which are included in the *What do I need to know?* section above.

Alternatively, the level and amount of adult or peer support can be varied according to the complexity of the activity. Activity 2 will probably require less adult intervention than Activity 3; hence you could arrange for more assistance for the activities which are more likely to require it.

Varying the level of challenge

Those children who are capable of more demanding work could be encouraged to pursue a more complex enquiry requiring the collection and analysis of more data. The football teams/colours investigation shown above is more complex than the initial colours/gender activity because there are more variables involved. When the pairs are deciding on their investigation between Activity 1 and Activity 2 you could ask the children to write their question or hypothesis down and you could sift through them to determine which pairs will need to be directed to more or less challenging variations on their intended investigation.

Why teach this?

The primary focus for this project is clearly the gathering and analysis of data and hence covers all statements in the ICT programme of study for 'Finding things out'. However, it also provides the children with an opportunity for 'Exchanging and sharing information' through the presentation of their findings through a word-processed report. In the final activity, the children have the opportunity to explain to others what they have done and use the graphs to back up their reasoning. The emphasis on the production of the reports should therefore be on making sure the children have taken account of others when reading their reports.

The focus for QCA ICT Unit 3A: *Combining text and graphics* is primarily making use of ICT to support aspects of English. However, combining text and graphics is something which is of value for any subject. The skills involved in Unit 3A include searching and replacing text and improving the quality of written work. Activity 4 could be adjusted to focus on improving the children's reports through editing and the use of spell checkers and an in-built thesaurus. This would enable the project to address the same ICT learning objectives as Unit 3A.

The ICT learning objectives for ICT Unit 3C: *Introduction to databases* are all covered by this project. In addition, because the focus is looking for relationships between two variables, this project is mathematically more demanding than Unit 3C.

Activity 4 of this project covers all the ICT learning objectives for ICT Unit 4A: *Writing for different audiences* by focusing on the needs of the audience in writing the report. The extent to which it addresses the English or Literacy objectives implicit in Unit 4A will be dependent on the timing for the project and the stage of development of the children.

Many of the learning objectives for Unit 4D: *Collecting and presenting information: questionnaires and pie charts* are covered by this project. Unit 4D makes use of pie charts to allow for comparisons between samples of different sizes. Activities 3 and 4 could be adapted to use pie charts for the comparisons, but given the small sample sizes this is not essential. A useful compromise might be to present the data with both pie charts and bar charts and ask the children to evaluate which they feel are the most helpful in enabling them to answer their questions. This will add an extra level of complexity to the project which will require professional judgement based on your knowledge of the children.

Following the data-handling cycle not only enables children to appreciate the way in which questions can be answered through the gathering and analysis of information; it also enables them to appreciate that mathematics is a means by which ideas can be tested. Quantification is the process by which numbers are used to explain aspects of everyday experience and hence embeds numbers into meaningful contexts for the children. A graph is a means of turning numbers into pictures – and, as we know, a picture is worth a thousand words!

The NNS focuses on the interpretation of data, which forms part of the data-handling cycle. The size of sample which the children will be using for these investigations will not require the use of scales grouped in twos, fives or tens, but this project could be followed up by looking at graphs of information drawn from the internet, magazines or newspapers. Having gone through the data-handling cycle, the children should have a greater understanding of how the data were gathered and the significance of the information in other people's graphs.

See also *Science* Project 4 (*Branching databases*), *Science* Project 5 (*Graphical representation of data*) and *Humanities* Project 6 (*Using a database to analyse census data*) for related activities.

Project Fact Card: Project 7: LOGO challenges

Who is it for?

- 8- to 10-year-olds (NC Levels 2–4)

What will the children do?

- Following an initial explorative activity with LOGO, the children will teach the turtle new procedures and then investigate how to teach it to draw a selection of shapes, graded for difficulty. Finally, they will apply their knowledge to produce a scene using a library of procedures which you have prepared (or downloaded)

What should the children know already?

- Some knowledge and experience of programmable toys (e.g. Project 1 or QCA ICT Unit 2D)
- The names of different types of triangles
- A basic knowledge of co-ordinates

What do I need to know?

- Some basic LOGO commands
- How to create and use a new LOGO procedure
- How to save, load and print from LOGO
- How to create a library of procedures and install these on the children's computers
- How to install the *SuperLOGO* demonstration program

What resources will I need?

- A version of LOGO
- A challenges sheet (see the CD-ROM accompanying this book for an example)
- A set of procedures to draw objects in a scene

What will the children learn?

- Some basic LOGO commands
- How to create and use LOGO procedures to control a screen turtle
- About the properties of plane shapes
- How to define co-ordinates in four quadrants

How to challenge the more able

- Encourage them to attempt the more complex challenges
- Suggest designs requiring more complex programming

How to support the less able

- Prepare a challenges sheet which is less demanding
- Provide more support

Why teach this?

- It covers ICT NC KS2 PoS statements 2b, 2c
- It complements or replaces QCA ICT Scheme of Work Unit 4E
- It addresses Mathematics NC KS2 PoS statements Ma3, 1c–e, 1g, 1h, 2a–c, 3a, 3c, 4c
- It reinforces NNS Y4 Units Au/Su 5–6, Sp 5–6; Y5 Units Au/Su 8–10, Sp 5

LOGO challenges

What will the children do?

Activity 1: Introduction to LOGO

Dependent on the prior experience of the children, you might decide to begin this activity by showing them a programmable toy such as *Roamer*, *Pip* or *Bee-Bot*. Most of the children ought to have had some experience with using this type of device and it will serve as a useful precursor to showing them the screen turtle.

Show them the screen turtle. Some may already have used LOGO or an intermediate program such as *Terry the Turtle* (Kudlian), BlackCat *Turtle 2* or *BlackCat Logo*. Ask if they know or can work out what needs to be typed in to control the turtle. Demonstrate the effect of:

FORWARD (FD), BACK (BK), LEFT (LT), RIGHT (RT) and
CLEARSCREEN (CS)

NOTE: Some versions of LOGO use **CLEAN (CL)** or **CLEARGRAPHICS (CG)** instead of **CS**.

Set the children some initial challenges (these could be written on the board, flip chart or on a handout) such as:

⊙ Find out how tall and wide the screen is.

⊙ Can you draw a square?

⊙ Can you draw a rectangle?

⊙ Can you draw a triangle?

    Equilateral?

    Right-angled?

    Isosceles?

    Scalene?

Ask them to write down the commands needed to draw each shape. Note that these challenges are arranged in order of difficulty – you could decide to initially give the children the first three and then add the others as needed.

In pairs, let them investigate each of the challenges. If they need to check whether they have closed the shape, then show them **HIDETURTLE (HT)** and **SHOWTURTLE (ST)** which enable them to see the lines underneath the turtle.

In the plenary, share the children's findings with the rest of the class. For example, some may have turned left to draw their square, others turned right. Some will have drawn a square of 50 units, others 100, etc. Reassure them that with a challenge, any solution is acceptable – as long as it meets the challenge.

Activity 2: Teaching the turtle new words

Recap on the previous activity. Remind them they made a note of the commands needed to draw the shapes in the previous activity. Ask if anyone can tell you how to draw a square. Draw one and then explain how handy it would be if the turtle knew how to do this already. Type in the word **SQUARE** to demonstrate that the turtle does not know how to square. Show them how to define **SQUARE** procedure. Show them how to edit the procedure to change the size of the square.

Explain they are now going to teach the turtle how to do new things. Working in pairs, the children write procedures for some of the shapes they drew in the previous activity.

At some point, demonstrate how they can use **REPEAT** with their procedures to make swirl patterns (see *What do I need to know?*, below) and then let the children experiment with swirling their shapes.

Activity 3: Graded challenges

Introduce the challenges sheet to the children (see the CD-ROM accompanying this book for an example). If necessary, work through one of the challenges with the children, using their ideas as to how the shape could be drawn. Emphasise that making mistakes is part of the problem-solving process and how they can edit their attempts through a '*trial and improvement*' approach.

Monitor the children as they attempt the challenges and, if necessary, adjust the level of challenge by guiding them towards those which are more appropriate.

In the plenary, share solutions to the challenges – particularly if there are different solutions to the same challenge. (NOTE: Some of the possible solutions to the problems can be found on the CD-ROM that comes with this book.)

Activity 4: Drawing a scene

Demonstrate to the children the library of shape procedures they can use for their scene. If you feel it is appropriate, you could also demonstrate **SETPENCOLOR (SETPC)**, **SETPOS** and/or **SETPENWIDTH (SETPW)**. Explain that they are going to combine the shape procedures into a '*super-procedure*' called '*scene*'. You will have to explain how the **GOTO** procedure makes use of co-ordinates, drawing upon their knowledge of the screen dimensions covered in the first activity. Emphasise that their **SCENE** procedure should be tested after each new object is added to make the debugging process easier.

The children work in pairs to produce their scene which could be saved on the network or on a USB (Universal Serial Bus) drive for the plenary. At some point you could suggest that new procedures can be added if the children want other objects in their scene (e.g. **PLANE**).

Some children might want to continue working on their scenes for more than one session. At some point you could have a 'work in progress' sharing session. If your version of LOGO includes drawing tools (e.g. *Imagine LOGO, MicroWorlds* (LCSI)), the children could embellish their scenes or print them out for colouring by hand.

What should the children know already?

Some knowledge and experience of programmable toys (e.g. *Project 1* or QCA ICT Unit 2D)

It is not essential that the children have used programmable toys earlier in their schooling but it will enable them to visualise that the screen depicts an aerial view of a turtle and help them appreciate that the commands apply to the turtle's orientation (i.e. its left and right). Showing them a programmable turtle at the start of the first activity could refresh their memories or provide a reference point for those who have not had this experience.

The names of different types of triangles

A recap on the properties of Scalene, right-angled, isosceles and equilateral triangles will be helpful for the children at the start of this project.

A basic knowledge of co-ordinates

The LOGO screen uses co-ordinates in four quadrants. Prior experience of defining positions on a grid with co-ordinates would be useful, though the LOGO screen provides a useful means of introducing this.

What resources will I need?

A version of LOGO

Any version of LOGO will be appropriate for this project, though some are more user-friendly than others. *SuperLOGO* (Logotron) has been used for the examples, which has recently been superseded by *Imagine LOGO*. The screen environment has changed slightly but the key components are largely the same. All versions of LOGO use the same basic commands but the way procedures are edited may vary slightly.

A free version of *MSW LOGO* can be downloaded from: http://www.softronix.com/logo.html. This website also provides links to a range of resources and ideas for making use of LOGO in the classroom.

A challenges sheet

An example is available on the CD-ROM for this book.

The challenges take the form of shapes and designs graded for difficulty which the children can select. For example, a Level 1 challenge would be to create a procedure to draw a rectangle. A Level 2 challenge would be to create a procedure to draw a hexagon. A Level 3 challenge would be to create a circle procedure. A Level 4 challenge would be to create a procedure to draw a sun (i.e. a circle with rays) and a Level 5 challenge would be to draw a square inside a circle.

Alternatively, the challenges built into *SuperLOGO* could be used:

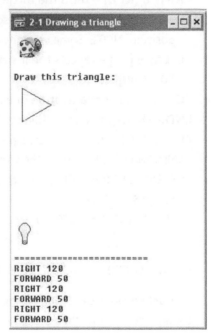

These are graded for difficulty and also include a solution. You should decide whether to let them view the 'solution' as it implies there is a 'correct' answer to the problem.

A set of procedures to draw objects in a scene

The subject for the scene could be up to you, but on the CD-ROM accompanying this book is a set of procedures for drawing a street scene – **HOUSE1**, **HOUSE2**, **TREE1**, **TREE2**, **CAR**, **SUN** and **CLOUD**. Also included is a procedure called **GOTO** *x y* which sends the turtle to co-ordinate (x, y) on the screen where (o, o) is the centre.

On the CD-ROM there are two sets of procedures. In the text folder is a set which should run on any version of LOGO. Copy and paste the procedures into the edit screen of your version of LOGO. There is also a *SuperLOGO Project* file which will run on *SuperLOGO* or *Imagine LOGO*. In addition, there is a demonstration program for *SuperLOGO* which can be installed on any computer to run the procedures.

What do I need to know?

Some basic LOGO commands

The commands which you and the children will need for this activity are:

FORWARD *n* (FD *n*) – moves the turtle forward *n* units, e.g. **FD 50**
BACK *n* (BK *n*) – moves the turtle back *n* units, e.g. **BK 40**
LEFT *n* (LT *n*) – turns the turtle left through *n* degrees, e.g. **LT 90**
RIGHT *n* (RT *n*) – turns the turtle right through *n* degrees, e.g. **RT 45**
CLEARSCREEN (CS) – clears the screen and returns the turtle to the home (centre)
    position. NOTE: Some versions of LOGO use **CLEAN (CL)** or **CLEARGRAPHICS (CG)**
REPEAT *n* [. . .] – repeats the instructions in the brackets *n* times, e.g. **REPEAT 4 [FD
    50 RT 90]**
TO xxx – starts the creation of a new procedure called xxx, e.g. **TO SQUARE**
END – shows the end of the list of instructions for a new procedure
PENUP (PU) – lifts the turtle's pen so it can move without drawing a line
PENDOWN (PD) – lowers the turtle's pen so it can draw
SETPENCOLOR *n* (SETPC *n*) – changes the colour of any lines drawn to colour code
    n, e.g. **SETPC 12**
SETPENWIDTH *n* (SETPW *n*) – changes the width of drawn lines to *n*, e.g. **SETPW 3**

How to create and use a new LOGO procedure

In most versions of LOGO, typing **TO xxx** will enable you to define a new procedure called xxx. To show that the list of instructions has ended, type the word **END**. For example, this will define a new procedure called **SQUARE**:

TO SQUARE
REPEAT 4 [FD 50 RT 90]
END

The way in which procedures are edited varies slightly from one version of LOGO to another. With some versions, typing in **EDIT** or **EDIT "SQUARE** will take you to the edit screen. Other versions of LOGO access procedure editing via a menu or by clicking on a button. In *SuperLOGO*, for example, procedures are edited by clicking on the **Memory** button:

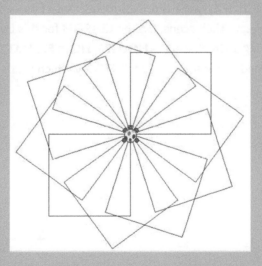

This will show a list of the procedures currently in the program's memory. Double clicking on the relevant procedure will allow you to make changes.

New procedures can be created in the edit screen. In *SuperLOGO* clicking on the **Objects** menu and then on **Add Procedure . . .** will allow you to do this.

To swirl the square:

REPEAT 10 [SQUARE RT 36]

Changing the number of repeats and the angle of turn will produce interesting variations. NOTE: The relationship between the repeat and the angle (i.e. 10 × 36 = 360). The turn of 36° is repeated ten times to bring the turtle full circle.

How to save, load and print from LOGO

Again, this varies across different versions of LOGO but loading, saving and printing is usually achieved by accessing the **File** menu.

For example, the file menu in *SuperLOGO* includes:

Saving a project will save all the procedures which are in the program's memory. Loading a project will load all those procedures. *SuperLOGO* allows you to print out the graphics screen and the text screen separately and will also allow you to print out the procedures you have written (**Print memory**).

How to create a library of procedures and install these on the children's computers

To complete the final activity you need to produce a set of procedures which the children can use to create their scene. On the CD-ROM for this book is a set of procedures for drawing a street scene – **HOUSE1, HOUSE2, TREE1, TREE2, CAR, SUN** and **CLOUD** – and a procedure called **GOTO** *x y* which sends the turtle to co-ordinate (x, y).

For example, the **SCENE** procedure for this screen is:

```
TO SCENE       GOTO 450 0
GOTO −300 0    TREE1
HOUSE1         GOTO 250 0
```

```
GOTO −100 0        TREE2
HOUSE1             GOTO −200 200
GOTO 100 0         CLOUD
HOUSE2             GOTO 200 200
GOTO 300 0         CLOUD
HOUSE1             GOTO −100 −100
GOTO −350 0        CAR
TREE1              GOTO −400 250
GOTO −150 0        SUN
TREE2              END
```

When creating their scene procedures you should advise the children to check the position of each new object before moving on to the next. This will make the debugging process a lot easier.

To copy these procedures into your version of LOGO
Open the text files in the **Resources** folder in the **Project 07** folder on the CD-ROM and copy and paste the text into the text screen or the edit screen of your version of LOGO. Pressing the **ENTER** key should define each procedure.

Once all the procedures have been installed and checked, save them as a project file for your version of LOGO.

How to install a *SuperLOGO* demonstration program

The scenery demonstration program is on the CD-ROM that comes with this book and can be installed on any computer free of charge.

To install the program open the **Project 07** folder on the CD-ROM, then double click on the **Install.exe** icon in the **Scenery install** folder which is in the **Resources** folder.

The program will be installed in the **Start > Programs** menu. Load the *SuperLOGO* demo and then click on the **demonstration files** folder (see Project 4) and then click on the **Scenery** icon.

To view the completed scene, type in *scene* and press the **ENTER** key.

What will the children learn?

Some basic LOGO commands

With very few LOGO commands the children can achieve a great deal. The commands are largely intuitive and hence easy to learn.

How to create and use LOGO procedures to control a screen turtle

Being able to create procedures unlocks the greatest potential of LOGO. Controlling the turtle directly by typing in commands can be very frustrating as there is no 'undo' facility. Being able to edit a procedure until the turtle performs the desired action puts the children more in control. For this reason, the scene created in the final activity should be created as a procedure, to enable the children to make amendments. Anticipating the outcomes of the instructions and then making use of the feedback received when the procedure is run means the children are processing the information at a deeper level and hence learning will be more secure.

About the properties of plane shapes

The first activity provides a means of investigating the extent to which the children are able to differentiate triangles and could be extended to explore regular polygons through the use of the repeat command (e.g. **REPEAT 6 [FD 50 RT 60]**).

How to define co-ordinates in four quadrants

If the **GOTO** procedure is used to draw the scene in the final activity, the children will be reinforcing their knowledge and understanding of the use of co-ordinates. Investigating the size of the drawing screen in the first activity will help them appreciate the maximum values which can be entered.

Challenging the more able and supporting the less able: modifying the project for older and younger pupils

Adjusting the level of challenge in the tasks

The potential of LOGO is immense – to the extent that it can be the subject of study at undergraduate level. The challenges sheet (an example is provided on the CD-ROM) enables the children to select those challenges which they feel are appropriate for them. The sheet could be modified to raise the levels of challenge for your children and you could intervene to increase (or lower) the level of challenge for specific pairs. For example, if a pair has written a procedure to draw a small circle, you could ask them if they could modify it to make the circle larger or suggest they try the concentric circle challenge (Level 5).

Some children will inevitably select challenges from the sheet which are inappropriate. You could suggest that all the children begin with a Level 1 challenge, or could differentiate by assigning levels for particular pairs.

The suggestions you make in the final activity should be geared to take account of each pair's capabilities. Those needing more support could be guided to placing objects at regular intervals. Those needing greater challenge might be asked to create a sub-procedure to draw, for example, an aeroplane, or shown how to colour the background.

Adjusting the level of support provided

Your knowledge of the children will enable you to identify those who are likely to need the greatest level of support. You may find that additional support would be helpful for the first two activities.

Why teach this?

This project develops the children's understanding of how ICT can be used to control events. LOGO is a fully developed computer programming language which was devised by Seymour Papert and his co-workers to provide a computer-based environment in which to learn through constructivist principles. Programming the turtle to solve shape and space problems not only provides a direct link between numbers and shape; it also enables the children (and you) to see immediately the effect of their understanding. The children do not need to be told by another that their solution to a problem is inappropriate; they can see on screen where their thinking is flawed and have the means to change their solution and check its validity. Thus, LOGO puts the children in control of the computer and in control of their own learning.

The structure for this project is similar to QCA ICT Unit 4E: *Modelling effects on screen.* It addresses the same learning objectives but raises the level of challenge and introduces more mathematically related concepts.

The mathematical focus for the project is problem solving with shape and space. Understanding of the properties of familiar shapes is used to help solve the challenges. The challenges provided on the example worksheet require the children to think about the relationship between combinations of shapes and develop their knowledge of angular and linear measurement. With LOGO, many of the turns require an appreciation of the relationship between internal and external angles and an intuitive understanding of the relationship between a diagonal and perpendicular lines. The final activity provides a natural context for the introduction or reinforcement of co-ordinates in four quadrants.

See also *Arts* Project 8 (*LOGO animation*) and *Arts* Project 9 (*Controlling external devices*) for related activities.

Project Fact Card: Project 8: Modelling investigations

Who is it for?

- 8- to 10-year-olds (NC Levels 3–4)

What will the children do?

- After exploring some online problems and investigations, and learning how to be more systematic in changing variables, the children will devise simple problems for each other using a spreadsheet

What should the children know already?

- How to load programs and access websites

What do I need to know?

- How to access and download activities from the internet
- How to use the online simulations
- How to enter data, labels and simple formulae into a spreadsheet
- How to format the appearance of a spreadsheet
- How to enter conditional functions into a spreadsheet

What resources will I need?

- Access to the internet and/or downloaded simulation activities
- Record sheet(s)
- A spreadsheet

What will the children learn?

- How to explore simulations systematically
- That simulations enable people to investigate 'What if . . .?' situations
- How to create and use a spreadsheet model
- How to use a spreadsheet to investigate

How to challenge the more able

- Explore more challenging simulations
- Create investigation spreadsheets with three (or more) variables
- Let the children experiment with their own spreadsheets

How to support the less able

- Restrict the simulations they explore to those which are less challenging
- Provide them with a template spreadsheet to modify
- Provide more support

Why teach this?

- It covers ICT NC KS2 PoS statement 2c, (2b)
- It complements QCA ICT Scheme of Work Unit 5D and/or augments Units 5E, 6B and 6C
- It addresses Mathematics NC KS2 PoS statements Ma2, 1a, 1d, 1f, 1k, 3c, 3d
- It reinforces NNS Y4 Units Au/Su 2, 8, 9, 12, Sp 8

Modelling investigations

What will the children do?

Activity 1: Using an online simulation

Select one of the simulations to use with the whole class. The Big Bus *Rocket* simulation is a good choice as it is visually appealing and the controls are easy to see and manipulate. Fire the rocket with the default settings and ask the children to suggest what might need to be changed to improve its flight.

If the children are having difficulty in solving the problem, suggest they change only one variable at a time to be more systematic in their experimentation. If, by chance, they solve the problem, then go back and explore the effects of changing only one variable at a time. Ask the children to predict what will happen if only one variable is changed.

Demonstrate the simulation which the children will explore. Explain how it can be accessed or loaded and then ask them to suggest how the controls might work. Suggest they explore the effects of the different variables.

In pairs, the children experiment with one or more of the simulations listed in the resources section. Prepare recording sheets for the children to log the values of the variables and the outcome (see the CD-ROM that comes with this book for examples). This could be done for each attempt or only for the solution – particularly for those simulations which have more than one successful combination of variables.

In a plenary session, discuss with the children:

⊙ what they noticed about the effect of the variables;

⊙ how realistic the simulations are.

If appropriate, you could ask some of the children to demonstrate to the rest of the class their solution – particularly if there is more than one successful combination of variables.

Explain how simulations or computer models are used by scientists, designers, architects and mathematicians to explore 'What if . . .?' situations. For example, before building a bridge, designers will use a computer to test the cost and effect of using different materials and to check its safety.

Use the *West Point Bridge Designer* simulation to demonstrate how changing the size of one of the beams in a bridge will reduce the cost but could prove disastrous.

Activity 2: Target practice

Show the class an example of the target practice spreadsheet file which you have created (see *What do I need to know?* and the CD-ROM accompanying this book for an example). Ask if they can explain how it might work. Tell them they are going to create a similar game for their friends to play.

When it comes to putting in the formula, the children need to select their own. To keep the mathematics straightforward the formula should not include division and you could restrict it to no more than three operations, e.g. $B3 * 3 - C3 + 2$. You could encourage some children to use brackets, e.g. $(B3 + 2) * (C3 - 1)$. After they have entered the formula and copied it to the lower cells, they should enter some values and enter each target. They should print out or record the values on paper and then delete them from the spreadsheet before saving it.

In a plenary session discuss how they managed the task and what they have learned about spreadsheets.

NOTE: You now have an opportunity to check through each pair's spreadsheet to ensure it functions properly before the next activity.

Activity 3: Solving each other's problems

The activity is introduced and explained to the class. The children may need to be reminded about being systematic.

Each pair loads its spreadsheets from the previous activity and then exchanges with another pair to solve the other's problem. When they have completed the task they print out the solution as evidence. If there is time, the children could then change the formula on their own or the other pair's spreadsheet and see if they can now reach the original targets but with the different formula.

In a plenary session, the children share with the rest of the class how well they managed to solve the problems and any strategies they developed. For example, once the children had worked out a few answers, could they use this information to help solve the others?

Activity 4: Some practice

Demonstrate the arithmetic practice activity the children will be creating (see *What do I need to know?*, below) and ask if they can work out how it knows the answers. Show them the 'IF' formula and explain and demonstrate its function.

Use the most appropriate method of instruction for guiding the children through the construction of their own spreadsheets (see Activity 2).

When they have completed the spreadsheet they should check it to make sure it is working properly and then delete the answers before saving it.

If there is time, pairs could swap to try out each other's practice tests.

In a plenary session, discuss how the spreadsheets work and ask if any of the children have suggestions for other ways in which the tests could be used. Explain and/or show them that the 'IF' statement can be used to check for words as well as numbers. For example,

=IF(B5='Africa','Correct','Incorrect')

Suggest that some of the children could follow up the activity at home by creating tests relevant to the work they are doing in other subjects.

What should the children know already?

How to load programs and access websites

Although not essential, you might find it useful if the children know how to access websites by typing the address (URL) into the address pane of a web browser so that they can move on to the next simulation during Activity 1. However, this could be taught within the context of the activity.

What resources will I need?

Access to the internet and/or downloaded simulation activities

Online simulation activities

The following websites provide online activities for the children appropriate for this project. The children will need to use computers which are connected to the internet for each of these activities.

⊙ *Rocket simulation (Big Bus)*

In this simulation, the children need to change three variables to control the flight of a pneumatically powered rocket.

http://www.thebigbus.com/activities/demo/rockets2.html

⊙ *Colin's coffee simulation*

In this simulation, the children must manipulate four variables to make Colin a perfect cup of coffee. After the initial problem has been solved a 'random' problem can be selected.

http://ngfl.northumberland.gov.uk/ict/qca/ks2/unit3D/colins%20coffee/colins%20coffee.html

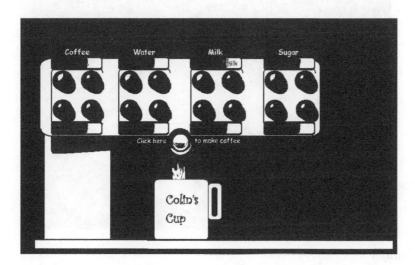

⊙ *Parachute game*

In this activity, the children need to change two variables to design the safest parachute. This is a useful activity as the level of challenge is increased as each problem is solved.

http://puzzling.caret.cam.ac.uk/game.php?game=parachute

Downloadable simulations which can be installed on computers without internet access

The following simulations can be downloaded from the internet (free of charge) and installed on any computer regardless of whether it is connected to the internet.

◉ *Balloon car*

This simulation requires the children to change four variables to design the most efficient balloon-powered car. Visually appealing but the children must record the results of their experimentation.

http://pbskids.org/zoom/games/ballooncar/

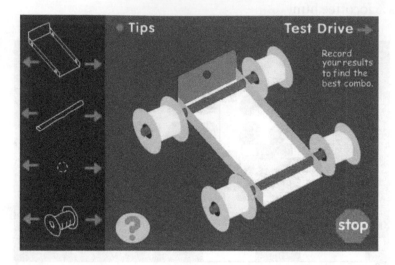

◉ *Duck Builder*

This simulation requires the children to adjust seven variables to create the ideal duck for flying. There are several successful combinations of variables so it would be useful if they recorded their successful solution for later comparison. Alternatively, you could vary the task by suggesting they set some of the variables to particular values and then explore how the other variables can be changed to take account of these.

http://www.cgpbooks.co.uk/online_rev/duck/duck.htm

⊙ *West Point Bridge Designer*

This program is too complex for the children to explore in this project but could be used in the plenary of the first activity to show how architects use ICT to help test their designs – if necessary, to destruction. The program can be downloaded free from

http://bridgecontest.usma.edu/

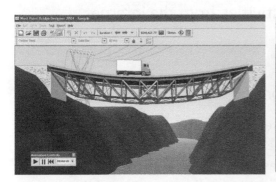 

NOTE: There are many more simulations available on the internet which could be used as alternatives to these. However, these are most appropriate for this project as the manipulation of the variables is prominent. Your choice of simulations could relate to the work they are doing in other subjects.

Record sheet(s)

For the children to note down the values of the variables they have set, and the outcomes. For example, a sheet for the *Duck game* could look like this:

Duck game

Variable Attempt	1	2	3	4	5	6	7	8	9	10
Front leg										
Back leg										
Front wing										
Back wing										
Body										
Bill										
Flap speed										
Outcome (Crash, Orbit, Fly)										

A spreadsheet

Any spreadsheet can be used for this activity, though if you would like the children to learn how to format the spreadsheet to improve the appearance of the activities they have developed, then you should use one of the more advanced spreadsheets such as *MS Excel*, BlackCat *NumberBox 2* or *Textease Spreadsheet*. Other educational spreadsheets provide a limited range of tools for enhancing the format and appearance of the completed sheet while some are very basic in the features they offer.

Enhancing the appearance of the completed spreadsheet is not essential to the project but does add an extra dimension and encourages the children to take a pride in their work.

What do I need to know?

How to access and download activities from the internet

See Project 3 (*Shopping*) for guidance.

How to use the online simulations

Most of the simulations mentioned here are intuitive in their use, being designed primarily for young children. A few minutes' experimentation will be sufficient for you and the children to grasp the basic principles. A simple sheet encouraging the children to record the values they have set for the variables and the outcomes of those settings would help to emphasise the importance of a systematic approach.

The *West Point Bridge Designer* program used in Activity 1 to demonstrate how simulations are used in the real world is slightly more complex and a little daunting when first loaded. However, to show the children how it works, load one of the example bridges (**File** > **Load sample design** > **Pratt Deck Truss (44 meter span)**). Once the bridge has loaded, test it with a load (**Test** > **Load test**). Return to the **Drawing Board** (**Test** > **Drawing Board**) and change the thickness of one or two of the middle beams by clicking on it to highlight it then clicking repeatedly on the **Decrease member size** button.

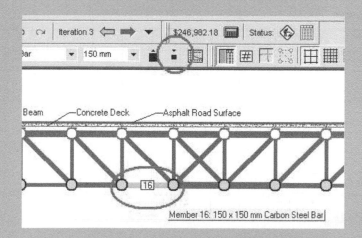

Note that as you are doing this the dimensions of the member decrease as does the overall cost of the bridge. Retest the bridge. Repeat changing the size of members as necessary.

NOTE: When closing the program DO NOT save the changes you have made to the example bridge otherwise the next time you demonstrate the program the bridge will collapse.

How to enter data, labels and simple formulae into a spreadsheet

The example shown here uses *MS Excel*, but the principles will apply to most educational spreadsheets such as BlackCat *NumberBox 2*.

1. Enter labels
Click on a cell and type in the text required:

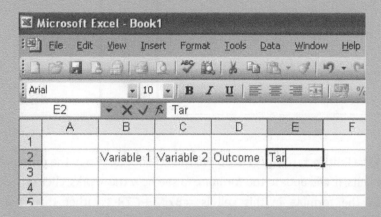

2. Enter a formula
Click on the cell in which you want the result of the calculation to be shown (in this example it will be cell D3) and then type in the formula. Note, a formula in a spreadsheet always starts with an equals sign (=). The example shown here (=B3*2+C3+1) means *take what is in cell B3 and multiply it by 2, then add what is in cell C3 and then add one*. The result of this calculation will be shown in cell D3, the cell containing the formula.

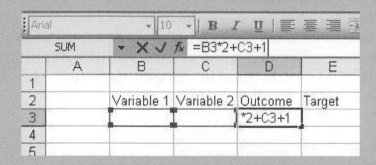

Press the **ENTER** or **RETURN** (↵) key to confirm the formula (or click on the green tick) and enter it into the cell.

At this point you might want to check that the formula works by typing some numbers into cells B3 and C3.

3. Copy the formula to the next nine cells down
There are at least two ways this can be done. It is useful to know both ways because of the variations in the way educational spreadsheets work.

Fill down

Click on the cell containing the formula, then drag down nine more cells (in this case from cell D3 to D12):

	A	B	C	D	E
1					
2		Variable 1	Variable 2	Outcome	Target
3				1	
4					
5					
6					
7					
8					
9					
10					
11					
12					
13					

Then from the **Edit** menu, select **Fill down**. This will not just copy the formula into the other nine cells, it will update the formula. To see how this works, click on cell D7 and notice that the formula in this cell is **=B7*2+C7+1** rather than the original **=B3*2+C3+1**.

Copy and paste

Click on the cell containing the formula you want to copy. Select **Copy** from the **Edit** menu. Then drag over all the cells you want it pasted into to highlight them (i.e. from D4 to D12) and then select **Paste** from the **Edit** menu. With some of the more basic spreadsheets you may not be able to highlight a series of cells and so may have to paste the formula into each one separately.

4. Change the font and text style

Highlight the cells in which you want to change the appearance of the text by dragging over them (in this case, all the cells from A1 across to E12). Then click on the **Format** menu and select **Cells . . .** Then click on the **Font** tab and make the required changes. In this case **Comic Sans, Size 16, Bold style.** Then click on the **OK** button to confirm these changes.

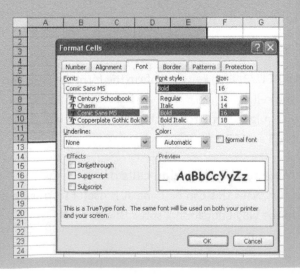

5. Changing the width of the cells

You may find now that the text doesn't fit into the original narrow cells. To widen the cells, move the pointer to the divider between the cells at the top of the window and drag to the new width.

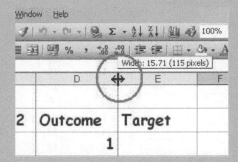

On some spreadsheets you may have to format the cell width through the **Format > Cells . . .** menu and on others double clicking in the grey area will automatically set the width of the column to the contents.

To create the target numbers, enter some values in the Variables columns and then type in outcomes into the Target column. At this point you could print out the spread-sheet (showing the answers) before deleting the answers and saving the spreadsheet.

How to format the appearance of a spreadsheet

1. To change the background and cell borders

Again, highlight all the cells you want to change and then click on the **Format > Cells > Pattern** and select the background colour you want for the highlighted cells.

To change the borders of the cells, highlight the cells you want to change and then click on **Format > Cells > Border** and click on the required style of border required.

2. To lock the cells

First, highlight any cells you do not want locked (i.e. the ones the children will type their variables into). Then select **Format > Cells > Protection**. In this dialogue box, remove the tick in the **Locked** box by clicking on it. This will ensure these cells will not be locked when the spreadsheet is protected.

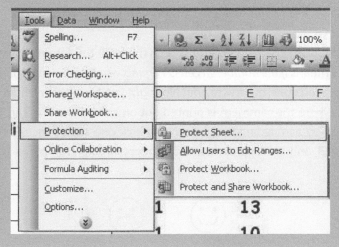

To lock the spreadsheet, click on **Tools > Protection > Protect sheet**. (NOTE: It is advisable not to enter a password unless you are sure it will be remembered.) Your completed spreadsheet should look something like this:

Note that some Clip Art has been added to enhance its appearance.

How to enter conditional functions into a spreadsheet

The spreadsheet for the final activity looks like this:

The function which is entered into cell H4 is:

=IF(G4=C4*E4, "Well done", "Not right yet")

The use of the brackets, commas and speech marks in the formula is very important. This formula means *if the number in cell G4 equals what is in C4 multiplied by what is in E4, then print **Well done**, otherwise print **Not right yet***. Note the formula is put into cell H4 which is where the comment is printed out on screen. The text can be changed, or even replaced by a Wingding image such as a tick or a smiley face, if the font in cell H4 is changed and the appropriate letter representing the image is put into the formula.

If you experience difficulty in typing an equals sign into a cell (the computer will think you are trying to enter a formula), type the equals sign and then press the enter key rather than clicking on another cell.

What will the children learn?

How to explore simulations systematically

Children's (and many adults') initial response when given a simulation is to enter variables in a haphazard way. In your demonstration, your interventions and also in the plenary for the first activity, try to emphasise the importance of being more systematic. Keeping all other variables constant while changing one and seeing its effects is a useful strategy, as is changing a variable in a structured way (e.g. in steps of 5).

That simulations enable people to investigate 'What if . . .?' situations

The simulations the children will use are fairly trivial, which is why showing them the bridge-building software in the plenary for Activity 1 is valuable in helping them appreciate how computer modelling is used in the real world. What if the client wanted a cheaper bridge? Could the materials be changed to cut costs? The way the bridge responds under the load test has deliberately been exaggerated in the program, to show the stresses in the members.

How to create and use a spreadsheet model

Spreadsheets are very versatile tools and knowing the basics of how they function is useful background knowledge for the children. A quick trawl of the internet will show that spreadsheets are used extensively by businesses to log and forecast expenditure.

How to use a spreadsheet to investigate

The first spreadsheet activity demonstrates how computers can be used to give immediate feedback. The difference between the simple investigation the children are using and the simulations they explored previously is that the formulae are more complex and the outcomes are translated into animations or text messages. The second spreadsheet activity shows them how feedback can be changed dependent on the input the computer receives. This lays the foundation for aspects of control technology.

The QCA does not introduce conditional statements until Year 6 (Unit 6C) but if the context is right and the activity is carefully taught, the concept is not too difficult for younger children to grasp.

Challenging the more able and supporting the less able: modifying the project for older and younger pupils

Choosing simulations which are suited to the needs of the children

The *Parachute game* and *Colin's coffee* simulations are motivating and are relatively easy to use. The *Balloon car* simulation is simple to use but the children will need to record their modifications and the distance travelled; hence this is a more complex simulation to explore. The *Duck Builder game* has the most variables out of these examples, but actually has a range of combinations to produce a successful outcome and hence is relatively easy to use.

More complex simulations than those listed include:

⊙ *Alien Plants* – http://www.mathsonline.co.uk/nonmembers/gamesroom/sims/plants/data.html

Although there are only three variables, the children will need to be persistent and well organised to determine the correct combination (don't forget decimals!).

⊙ *Planet 10* – http://www.planet-science.com/wired/?page=/planet10/

The children need to keep track of a large number of variables to design their own planet and then be prepared to monitor the results of their handiwork and take account of the feedback to make modifications.

⊙ *Plant Force* – http://puzzling.caret.cam.ac.uk/game.php?game=plants

Children have to adjust the temperature and water given to a plant on a weekly basis and monitor its growth and monetary value to make a profit. This simulation can be frustrating as the effect of changing the variables can be cumulative.

Two more simulations which could be used as extension tasks, but have less mathematical learning potential than the previous examples, are:

⊙ *Funderstanding Roller Coaster* – http://www.funderstanding.com/k12/coaster/

The variables are controlled via sliders and hence are easy to manipulate, but there is no numerical output. There are seven variables to control but there is no reason to change the mass and friction unless you direct the children to do so.

⊙ *Roller-coaster Designer* – http://puzzling.caret.cam.ac.uk/game.php?game=roller

Fun to play, but the control of the variables is more hit and miss meaning that the children are not made aware of the maths involved.

Varying the challenge with the spreadsheet tasks

For those lacking in experience or confidence with ICT, a template spreadsheet could be prepared which the children modify under your guidance.

To increase the challenge, you could ask some pairs to work out how to add another variable, or encourage them to make the formula more complex, and then see if they can reach the same target numbers with the changed formula.

Some children might be interested in developing either of the spreadsheets further as a homework task.

Varying the level of support

This can be achieved by using more adults, enlisting peer support (maybe from well-briefed children in an older class) or by creating step-by-step handbooks to guide the children through the processes involved.

If you feel the relationships in the class are appropriate, some of the more confident children could be used as peer 'monitors' to assist those with less experience of ICT.

Why teach this?

The first three activities directly address statement 2c of the KS2 Programme of Study for ICT, while Activities 2–4 show how spreadsheets can be used as a tool to assist with other aspects of the children's work. The use of the conditional function (IF) in Activity 4 introduces them to aspects of control. Computers have the capability of interacting with their surroundings and making limited decisions about what to do in certain circumstances. In this case, the computer is monitoring the value which is entered into a cell and choosing to make one of two responses. Although this may seem to be an apparently insignificant function, it is actually one of the most powerful features of computer technology.

Many of the ICT learning objectives covered by QCA ICT Unit 5D: *Introduction to spreadsheets* are addressed in this project. This project could be combined with activities in ICT Unit 5D and/or replace them. The party-planning activity in Unit 5D could replace Activity 2 or Activity 4. However, the mathematical aspect of the party-planning activity tends to become buried in the complexity of the development of the activity and the opportunity to introduce the conditional function would be lost.

Having some experience of the conditional function in this project would act as valuable prior learning for Units 5E: *Controlling devices* and 6C: *Control and monitoring* – 'What happens when . . .?'

The investigation into rectangular area in Unit 6B: *Spreadsheet modelling* could readily be combined with this project and/or act as an extension activity for those children who are ready to move on to more challenging tasks.

The simulations should be used to encourage the children to adopt a more systematic approach to manipulating the variables and to take account of the effects of their actions. The first spreadsheet activity is intended to encourage them to focus principally on the relationship between the numbers they are entering as variables and the outcome. As they work down the rows they should become more efficient at predicting the likely outcome of combinations of variables and some will probably identify the pattern.

The simulations in the first activity encourage the children to think about the relationship between mathematics and real-world situations. Activity 3 is designed to encourage them to apply the systematic approaches developed in Activity 1 to the solution of the simple number problems. The problems can readily be modified by changing and copying down the formula. Even if they know the formula they have entered, the mental calculation involved in working back from the target to the variables is valuable in practising their mental calculation skills.

The second spreadsheet task (Activity 4) could be used to help children improve their rapid recall of addition, subtraction and multiplication facts.

See also *Arts* Project 3 (*Designing an environment*) and *Science* Project 6 (*Giant's hand*) for related activities.

Project Fact Card: Project 9: Patterns and spreadsheets

Who is it for?

- 9- to 11-year-olds (NC Levels 3–5)

What do I need to know?

- How to access and/or download online activities
- How to create function machines and number crunchers with a spreadsheet
- The laws of commutativity and associativity
- How to generate number patterns

What will the children do?

- After exploring computer-based and online function machines, they will create their own using a spreadsheet. They will use this to explore the effect of changing the order of calculations. They will then modify their function machine to a 'number cruncher' to explore the number patterns arising from the matchstick patterns

What resources will I need?

- A computer-based function machine program
- Online function machines and number crunchers
- A spreadsheet program suitable for use in KS2 (e.g. BlackCat *NumberBox 2*).
- Function investigation worksheet
- Pattern investigation worksheet

What should the children know already?

- That there are four 'operations' (add, subtract, multiply and divide)
- The main features of a spreadsheet

What will the children learn?

- How to enter formulae into spreadsheets
- How to use spreadsheets to investigate functions and number patterns
- The interrelationships between number operations
- The foundations of algebra

How to challenge the more able

- Increase the level of mathematical challenge by investigating more demanding number patterns
- Let the children explore the effects of three number operations in the spreadsheet function

How to support the less able

- Scaffold the activities more carefully
- Provide more adult support

Why teach this?

- It covers ICT NC KS2 PoS statements 2b, 2c
- It complements QCA ICT Scheme of Work Units 5D and 6B
- It addresses Mathematics NC KS2 PoS statements Ma2, 1a–c, 1h, 1k, 2b, 3a, 3c, 4d
- It supports NNS Y5 Units Au/Su 12, Sp 11; Y6 Units Au/Su 12, Sp 11

Patterns and spreadsheets

What will the children do?

NOTE: See *What resources do I need?* (below) for information about the online resources used in the activities.

Activity 1: Investigating a computer-based or online function machine

Demonstrate a basic function machine to the children such as that provided by Ambleside School. Concentrate on developing a systematic approach to investigating the 'rule'; for example, entering 0 followed by 1, 2, 3, etc.

The children then work in pairs using the Shodor online *Function Machine*. Those needing more challenge could be moved on to the *Linear Function Machine*.

At the end of the activity discuss whether the children developed any strategies for finding the functions and whether any noticed anything interesting about relationships between input and output numbers (e.g. that when multiplying by 3 the output went up in threes).

Activity 2: Creating a simple function machine

Recap on the previous activity and explain that the children are going to devise their own function machines. Demonstrate the machine you have created. Initially suggest that the children use only one operation (e.g. + 9 or × 3).

Working in pairs, the children create their own function machines using a spreadsheet (see *What do I need to know?*). You could use one or a combination of three approaches for this:

⊙ You demonstrate the whole process and then the children create their own.

⊙ Present the process in chunks, the children completing each chunk before moving on to the next.

⊙ Create a manual for them to follow showing the process.

Once each pair has created their function machine, they swap with another pair to try and solve the other's problem.

Those requiring additional challenge could combine two operations (e.g. $+ 3 \times 2$). Alternatively, you could show them how to 'prettify' their function machines through the addition of Clip Art etc.

The pairs should save their function machines as they will be needed for the next activity.

In the concluding plenary, ask one or two to demonstrate their function machines for the whole class to try and solve – and ask if anyone has developed a 'fool proof' strategy for solving others' functions.

Activity 3: Creating number crunchers

Demonstrate the Shodor *Number Cruncher*. Discuss how having the results displayed might help them identify the hidden functions. Show how being systematic with the input numbers results in a pattern of output results which might assist in identifying the function.

Working in pairs, the children investigate one or both of the Shodor *Number Crunchers*.

Bring the children together and discuss what they found out. Show them how to modify their initial function machine spreadsheets to generate a list of results (see *What do I need to know?*).

Working in pairs, the children modify their function machine spreadsheets and then test them with different functions. They must save their number crunchers for the next activity.

In the final plenary, ask if anyone spotted any interesting patterns.

Activity 4: Investigating functions

Recap on the previous activity. Ask the children to predict if there will be a difference between the output of cell $* 3 + 2$ and cell $+ 2 * 3$. Use one of the children's (or your own) *Number Cruncher* to investigate the difference. Show them the 'Function investigation worksheet' (provided on the CD-ROM accompanying this book) and ask them to predict which pairs of functions they think will produce the same set of results and which will produce different results. They can then choose the ones they want to test using their number crunchers.

Some of the functions they can investigate are:

$* 3 + 2$ and $+ 2 * 3$ (i.e. changing the order of the calculation)
$* 3 + 2$ and $+ 3 * 2$ (i.e. changing the operation)
$* (3 + 2)$ and $(* 3 + 2)$ (i.e. using brackets)

The level of challenge could be increased by investigating combinations of three operations:

$+ 3 * 2 - 1$ and $(+ 3 * 2) - 1$ or $+ 3 * (2 - 1)$

113

Some children might want to investigate their own functions.

A format for the investigation sheet could be:

Names: _____   Date: _____

Number Cruncher Investigations

You are going to investigate whether these functions are different or just the same ones in disguise.
Use your detective skills to predict whether the outputs will be different or the same!

Function 1	Function 2	Prediction Diff / same	Checked Diff / same
+ 3 + 6	+ 6 + 3	same	same
+ 2 + 3	+ 3 + 2		
+ 9 – 1	– 1 + 9		1
– 7 + 4	+ 4 – 7		
+ 3 – 1	– 3 + 1		
+ 8 – 7	– 8 + 7		
* 3 + 2	+ 2 * 3		
+ 5 * 2	* 2 + 5		
* 0 + 9	+ 9 * 0		
* 3 – 4	– 4 * 3		
(* 2 + 1)	* (2 + 1)		
+ 5 * 3	+ (5 * 3)		

Now investigate some of your own ideas

You could ask the children to print out the results from each investigation – though this will involve heavy paper usage.

In the final plenary, discuss what the children have noticed. For example, they might have noticed that when the operations are the same (e.g. two adds) the output is the same when the order of operations changes, but when the operations are different (add and subtract) the outputs change.

Decide whether to formalise their observations by creating a set of 'rules' – e.g. 'When the operation is the same, changing the order doesn't matter'.

Activity 5: Finding functions from patterns

Explain to the children they are going to use their *Number Crunchers* to solve some more mysteries. Demonstrate the first investigation from the 'Pattern investigation worksheet' (provided on the CD-ROM accompanying this book) using an overhead projector and matches to show how the pattern develops.

Pattern 1

Number (Input)	Pattern	Sticks (Output)
1	L	2
2	LⵏΓ	4
3	L⎍Π⎍	
4		
5		

Function: Input _____ = Output

Pattern 2

Number (Input)	Pattern	Sticks (Output)
1	‾‾I	3
2	‾□I	
3		
4		
5		

Function: Input _____ = Output

Show them how to use their *Number Cruncher* to work out what function will produce the pattern of results in the output column.

Distribute the sheets and ask the children, in pairs, to work out the patterns and the functions.

In the plenary at the end of the lesson, discuss what the children noticed about the patterns. For example, they may have noticed that Patterns 3 and 4 were the same as Patterns 1 and 2, but with one (stick) added (i.e. * 2 + 1 and * 3 + 1).

What should the children know already?

That there are four 'operations' (add, subtract, multiply and divide)

This could be reinforced in the first activity. It might be that some children have not appreciated that these are the only four numerical operations and some may not be aware of the relationships between addition and (its inverse) subtraction, or (repeated) addition and multiplication, or (repeated) subtraction and division, or multiplication and (its inverse) division.

The main features of a spreadsheet

It is not essential that children have prior experience of spreadsheets, as an introduction to them could be incorporated into the first session, but it would ease the learning demands if the children had some experience with spreadsheets before engaging in this project.

What resources will I need?

A computer-based function machine program

Ambleside Function Machine

A useful program for the introductory whole-class discussion as the function is revealed without the need for checking. The program is downloadable free from:

http://www.amblesideprimary.com/ambleweb/mentalmaths/functionmachines.html

Alternatively, the function machine which is included with *Easiteach Maths* resources could be used.

Online function machines and number crunchers

Shodor Function Machine

This includes interactivity by asking the children to input the rule. It is an ideal first investigational activity as there is only one operation (add, subtract, multiply or divide) and one variable (e.g. + 5, × 3, − 4 etc.).

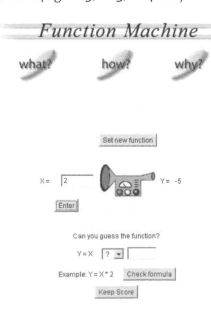

It is accessible online at:

http://www.shodor.org/master/interactivate/activities/fm/index.html

Shodor Number Cruncher

This is similar to the above function machine but provides the children with a list of their inputs and outputs. It provides a useful introduction to exploring number patterns (Activity 3).

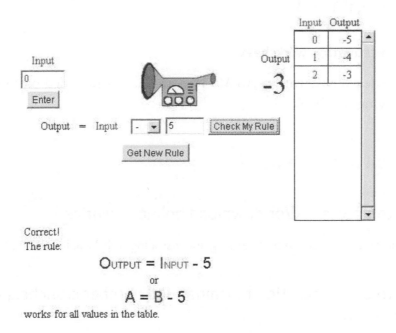

Access it online at:

http://www.shodor.org/master/interactivate/activities/numbercruncher/index.html

Whole number Cruncher *(Multiplication or addition)*

This is similar to the above but includes multiplication and addition operations. It is available online at:

http://www.shodor.org/master/interactivate/activities/pnumbercruncher/index.html

Linear Function Machine

This is a similar concept to the basic function machine but combines two operations (e.g. $\times$ 3 + 1):

http://www.shodor.org/master/interactivate/activities/lfm/index.html

A spreadsheet program suitable for use in KS2

In the examples below, *MS Excel* has been used, as this is probably the spreadsheet most universally available in school. Any educational spreadsheet can be used for this project as the level of sophistication required in the creation of the functions is not high. A disadvantage in using a simple spreadsheet might be that the children have less opportunity to enhance the appearance of their function machines through the use of colour, borders and Clip Art.

Function investigation worksheet

This is needed for Activity 4. An *MS Word* version is provided on the CD-ROM accompanying this book.

Pattern investigation worksheet

This is needed for Activity 5. An *MS Word* version is provided on the CD-ROM accompanying this book.

What do I need to know?

How to access and/or download online activities

See Project 4 for guidance on accessing and installing web-based activities.

How to create function machines and number crunchers with a spreadsheet

To create a function machine:
Open your spreadsheet program (in this example we are using *MS Excel*).

1. In cell B4 type **Input** and press the **ENTER** key.
2. In cell D4 type **Output** and press the **ENTER** key.
3. In cell D5 type in a formula such as =**B5+7** and press the **ENTER** key.
4. Format the cells to your preferred font, font size, background colours and borders and insert some appropriate Clip Art (see Project 8).

5. To change the function, change the formula in cell D5.

NOTE: The sign for multiplication in a spreadsheet formula is * and for division is /.

To create a number cruncher:

1. Create a function machine as above.
2. Click on cell B5 and drag down for another 11 cells.
3. From the **Edit** menu click **Fill down** (or **Copy** and **Paste** to the 11 cells below).
4. Click on cell D5 and drag down for another 11 cells.
5. From the **Edit** menu click **Fill down** (or **Copy** and **Paste** to the 11 cells below).
6. If necessary, change the formatting for these cells to something more appropriate (i.e. so that all the cells fit on the screen without the need for scrolling down).

7. To change the function, the formula in cell D5 will need to be changed and then this cell must be filled (or copied) down to the 11 cells beneath.

The laws of commutativity and associativity

The Commutative Law

For addition and multiplication, the order in which the calculation takes place makes no difference to the result: i.e. $a + b = b + a$ **or** $a \times b = b \times a$

The Associative Law

Put simply: $(a + b) + c = a + (b + c)$ **and** $(a \times b) \times c = a \times (b \times c)$

That is, when numbers are regrouped for addition or multiplication the result remains unchanged, i.e. $a + b$ followed by adding c gives the same result as $b + c$ followed by adding a.

Hopefully, though not essentially, these sorts of relationships will be noticed by children in Activity 4.

How to generate number patterns

The 'matchstick' patterns indicated above could be extended. For example:

	1	2	3	4	5	
▽△▽	3	5	7	9	11	input x 2 + 1 = output

	1	2	3	4	5	
⊠⊠	5	9	13	17	21	input x 4 + 1 = output

	1	2	3	4	5	
85	7	12	17	22	27	input x 5 + 2 = output

	1	2	3	4	5	
	12	22	32	42	52	input x 10 + 2 = output

Other patterns can be made with dots, squares or cubes. Consulting books and websites on number patterns will yield even more – though some fairly innocent-looking patterns are based on quite complex functions.

Here are a few number pattern websites:

⊙ *Active Algebra* (could be used as a follow-up activity) – http://www.active-maths.co.uk/algebra/sequences/matchsticks/

⊙ *NZ Maths:* lesson plan and more teaching ideas (non-computer) – http://www.nzmaths.co.nz/Algebra/Units/matchstickpatterns.htm

⊙ *Teaching Ideas* – http://www.teachingideas.co.uk/maths/nopattern/contents.htm

What will the children learn?

How to enter formulae into spreadsheets

How to use spreadsheets to investigate functions and number patterns

The project focuses principally on using functions to help them explore mathematical relationships. The process of creating formulae and functions will quickly slip into the background as the children become interested in the mathematics of the problems. Hence creating formulae is the means to an end, and not an end in itself.

The interrelationships between number operations

The fourth activity particularly makes use of the spreadsheet's ability to provide children with consistent and immediate feedback on their manipulation of combinations of operations.

The foundations of algebra

A spreadsheet formula is an algebraic function involving variables, input and output. By the end of the project the children will have learnt that a variable can be changed and the order in which algebraic expressions need to be calculated (i.e. BODMAS – *Brackets, Of, Division, Multiplication, Addition, Subtraction*). They should also appreciate the inverse relationships between multiplication and division, and addition and subtraction.

Challenging the more able and supporting the less able: modifying the project for older and younger pupils

Adjusting the mathematical demand of the patterns being investigated

The functions and matchstick patterns used in the project have been selected to provide a measured approach. However, some children may be ready to move on to far more challenging number patterns (e.g. square and maybe even triangular numbers).

Those requiring more support may need to have some of the activities (e.g. Activity 4) more carefully scaffolded. As indicated above, there are some interesting online activities which could complement the project.

Adjusting the level of support provided

You may find having the assistance of another adult helpful, particularly with the first and/or second activity. The ICT demands of the activities remain much the same and so you might find there is less need for support as the project progresses.

Why teach this?

During the course of the project the children are also learning about ICT and how it can be used as a tool to assist in trying things out. A spreadsheet formula is an instruction to the computer to do something. If the instruction is not phrased appropriately, the computer will either produce the wrong result or will communicate its confusion to the children. They will be continuously reviewing, evaluating and modifying their work as they progress through the project. At times they will need assistance, but they should also be encouraged to try out their own solutions before resorting to your 'expert' help.

QCA ICT Unit 5D: *Introduction to spreadsheets* focuses on the creation of a spreadsheet model to predict party expenditure. This provides the children with an interesting and purposeful context for making use of the spreadsheet's features. However, once the formulae have been entered, the focus shifts to the interpretation of the results and hence the children do not have the opportunity to engage with the effects of manipulating the formulae – a valuable mathematical and ICT experience. This project could be used alongside Unit 5D or could replace it.

The main activity in ICT Unit 6B: *Spreadsheet modelling* (exploring the relationship between the areas and perimeters of rectangles) could be used as a follow-up to this project. Unit 6B, as with Unit 5D, lays greater emphasis on interpreting the results of the investigation and although the spreadsheet is being used to assist in the process, the children do not really engage deeply with the effect of the formula.

Function machines have been used to help children explore number relationships for many years before ICT was introduced to the classroom. The notion of a 'function' acting as a 'black box' changing an input to an output is not only a useful metaphor for mathematics; it has parallels in computing and information processing.

The principal application of mathematics is the solution of problems. This project should help children recognise that patterns which exist in the world around them can be represented and explored mathematically, and that a computer package such as a spreadsheet can assist in the process by automating some of the laborious work. This process of automation enables the children to focus more on relationships between numbers and how arithmetical expressions can be adjusted to produce specific results.

The *National Numeracy Framework* places an emphasis on the solution of puzzles and problems and the expression of generalised relationships arising from specific cases. This project takes the children through this process by making use of the spreadsheet's facility for generating data quickly and accurately to help the children focus on patterns of results and shorthand ways of representing the relationship between one set of numbers and another through formulae. The children do not need an adult to tell them their formula is inappropriate: they will see immediately that it does not produce the required set of results and hence may need to make 'informed' adjustments until it does. Thus ICT is serving the needs of the learner in a mathematical context.

See also *Science* Project 5 (*Graphical representation of data*), *Science* Project 6 (*Giant's hand*) and *Arts* Project 7 (*Using a spreadsheet model*) for related activities.

Who is it for?

- 9- to 11-year-olds (NC Levels 3–5)

What will the children do?

- The children will explore a set of data produced by a class of 7-year-olds to investigate whether there is a link between physical features and athletic ability. They will then form their own hypotheses, gather data and analyse them to determine whether performance and physical features seem to change over time

What should the children know already?

- What a database is and how it works
- How to enter and analyse data in a database (or spreadsheet)

What do I need to know?

- How to conduct a statistical investigation
- How to import a file into a database
- How to set up a database
- How to present and interpret results using a scattergraph
- How to form and test hypotheses

What resources will I need?

- A data file of results from 7-year-olds' investigation
- An educational database (or spread-sheet) capable of drawing scatter-graphs (such as BlackCat *Information Workshop 2000*).

What will the children learn?

- How and why to use a database (or spreadsheet) to analyse data
- How to form and test hypotheses
- How to interpret scattergraphs
- How to conduct a statistical investigation

How to challenge the more able

- Encourage the children to form hypotheses which will require more careful analysis
- Follow up the initial investigation with one which is more focused

How to support the less able

- Scaffold the investigation to provide step-by-step support
- Provide more adult (or peer) support

Why teach this?

- It covers ICT NC KS2 PoS statements 1a–c
- It complements QCA ICT Scheme of Work Unit 5C and augments Unit 6D
- It addresses Mathematics NC KS2 PoS statements Ma4, 1a–c, 1e–h (some children), 2a–c, 2f
- It reinforces NNS Y5 Units Au/Su 6, Sp 8; Y6 Units Au/Su 6, Sp 8

Statistical investigations 2

What will the children do?

Activity 1: Analysing the Year 3's data

Show the children the data gathered by the Year 3 class which they used to test hypotheses such as 'Those with the longest legs can jump the furthest (or highest)'. Ask the children if they can work out how the database tools could be used to help them check these hypotheses. Demonstrate the sort and graph facility and also the scattergraph. You may need to help them interpret the scattergraph (e.g. by showing that each point represents a child). To demonstrate a strong correlation you could plot height against reach.

Ask the children if, given the data, they can think of any other possible correlations. Make a note of them on a whiteboard or flip chart.

Working in pairs, the children now investigate whether there are any strong correlations in the data. You could either give them a recording sheet, ask them to devise their own, or ask them to produce a report (see Project 6).

Share and discuss their findings in a plenary session.

Activity 2: Forming their own hypotheses

Recap on the previous activity and flag up any interesting findings which were made. Discuss whether the children think there may be any differences between the results found by the seven-year-olds and the results which a group of Year 6 (or Year 5) children might get. The most obvious will be that the children will have grown, but you could speculate on whether the relationships between height and reach might have changed and whether some of the hypotheses which the seven-year-olds explored might now have stronger correlations (e.g. that taller ten-year-olds might be better at high jump than shorter ten-year-olds). NOTE: You might prefer not to gather data about the children's weights.

The children split into small groups (of four) to form various hypotheses which they would like to explore.

Their hypotheses are shared in a plenary session and the methods of gathering data are explored. If the children are participating in athletics training (e.g. for sports day), then opportunities will be provided to gather performance data; otherwise classroom-based activities could be devised. For example, long-jump performance can be determined through a standing long jump in which children stand with both feet together on a line and jump. Standing high jump can be measured by marking a point on a wall which is their normal reach, smearing chalk on their fingers and jumping to touch the wall as high as they can. The jump is measured from their normal reach to the mark left by the chalk. Throwing performance would have to be measured outside unless an airflow ball is used in the hall.

At this point the children could also design the record format they will need for collecting the data on each child's performance and statistics. For example, this record has been created using *Textease Database*:

The fields required will be dependent on the hypotheses the children want to test.

The database could be set up in the plenary, with the children advising, or the fields could be listed and the database set up by the teacher or a pair of children before the next activity.

Activity 3: Gathering and entering the data

This next activity will need careful preparation to ensure that all the children have a clear understanding of their roles and responsibilities. Pairs of children should be allocated to supervise each data-gathering point and equipped with the relevant measuring equipment and a record sheet. In addition, each child should have a record card on which to write his/her own data. It is a good idea to allocate two pairs to each measuring station so that the original pair can be relieved at some point to enable them to participate in the data gathering.

The children then move round from one measuring point to the next until all the data have been gathered.

The way the data are entered into the database could be managed in one of several ways:

⊙ Each child enters his/her own data into the database. If you have a database which allows multiple entry via a network (e.g. *Textease Database CT*), then this can be done in the computer suite in one session. Alternatively, one computer could be set up in the classroom with the children taking turns to enter their data on a rota basis.

- ⊙ The measuring point supervisors could enter the data into the database. This can become laborious, however, which could lead to errors.

- ⊙ You or a colleague enters the data outside teaching time. Very time consuming but at least you will be able to check the results for anomalies.

Activity 4: Analysing the data to test their hypotheses

Brief the children again on the purpose of the activity and the process you want them to follow. You could ask them to produce a report showing the data from the Year 3 class and from their own investigation, for example. Stress the importance of presenting their findings so that others can make sense of them and that their reports will be shared at the end of the session via a classroom display (and/or via the class website).

Working in pairs, the children analyse the data to test their hypotheses and then write up a report of their findings.

In the plenary, discuss what the children found and if there are any other lines of enquiry which might have emerged. If you have time, these could lead to some follow-up investigations.

What should the children know already?

What a database is and how it works

Having some experience of databases (e.g. Project 6) will speed up the familiarisation process. If you are uncertain as to the children's past experiences with databases, you could spend more time in the first activity reinforcing the key features of a database (see Project 6).

How to enter and analyse data in a database (or spreadsheet)

This is not an essential pre-requisite for this project as data analysis and entry can be addressed in Activities 1 and 3. However, if the children are already familiar with the use of databases, the focus can be shifted on to analysis and interpretation rather than technicalities.

What resources will I need?

A data file of results from seven-year-olds' investigation

This is provided as a csv (comma separated variable) file, a tsv (tab separated variable) file, a text file (.txt), an *MSExcel* file (.xls) and a *Textease Database* file (.td) on the CD-ROM accompanying this book. At least one format should enable it to be imported into most database or spreadsheet programs.

An educational database (or spreadsheet) capable of drawing scattergraphs

Some (but not all) educational database and spreadsheet programs enable the children to compare two sets of data in a scattergraph (or scattergram). You should check to ensure this feature is provided on your program. Scattergraphs can be drawn on *Granada Database* (Granada Learning), *Information Workshop 2000* (BlackCat), *Information Magic* (RM), *Textease Database* (Softease), though this is not a definitive list.

What do I need to know?

How to conduct a statistical investigation

See Project 6 (*Statistical investigations 1*) for an outline of the statistical investigation process. This underpins the approach to this project and so a few minutes familiarising yourself with the principles and educational rationale for an enquiry-led approach will help focus your teaching on the key issues.

How to import a file into a database

The *Y3-stats* file provided on the book's CD-ROM is in five different formats which should enable you to import it into most databases. Each database will vary in the outward approach it uses to importing data, but the underlying process will be the same.

⊙ Load the *Textease Database* (**Start** > **Programs** > **Softease** > **Database**)
⊙ From the **File** menu select **Open . . .**
⊙ This will show the **Open file** dialogue box. From the **Type of file** drop-down list select **Comma separated (*.csv)**:

- Use the **Look in** box to locate where the file is saved on the CD-ROM (in the **Project 10** folder).
- Click on the **Y3-stats.csv** file and then click on the **Open** button.
- The file should now open and display down the page. To make the data easier to read, you could highlight the lower set of fields and drag these alongside the upper ones:

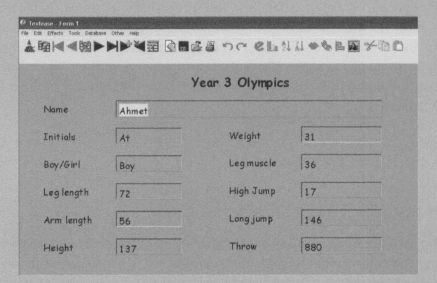

A *Textease Database* record

NOTE: With some databases or spreadsheets, you may need to click on **Import** or **Import data file** in the **File** menu to begin this process.

How to set up a database

Again, this process may vary in terms of fine detail from one program to the next, but the underlying principles will be the same. Refer to Project 6 for information on setting up your own database.

How to present and interpret results using a scattergraph

The procedure shown here is for the *Textease Database* but the process will be similar for other databases or spreadsheets.

- Let's assume we want to correlate **Arm length** with **Throw** to see if there is a relationship between these two (i.e. those with longer arms can throw further).
- Double click on the **Arm length** field and then right click on the **Throw** field so that both are highlighted.

⊙ Now click on the **Graph** icon on the menu bar at the top of the screen:

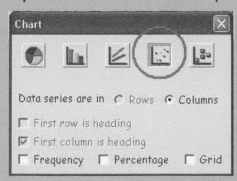

⊙ Now click on the **Scattergraph** icon to display the data as a scattergraph:

⊙ Other database programs do not require you to identify the two sets of data to be graphed until you reach the stage where you have chosen the scattergraph option. If in doubt, consult the manual or the on-screen help.

Interpreting scattergraphs

As with all graphs, scattergraphs require some understanding of what is being represented for the children to be able to interpret them. Probably the best way to explain what is being shown is to return to the question which prompted the use of the scattergraph: in this example, 'Is there a connection between the length of your leg and how high you can jump?' If there was a connection, we would assume that those with shorter legs would jump the least high, and those with the longest legs would jump the highest and, most importantly, all those in between would have jumps in the same proportion to the length of their legs.

On the scattergraph, each point represents a person. Polly has the shortest legs and so her point is the lowest on the 'Legs' axis. Jon has the longest legs and his point is the one highest on the 'Legs' axis. If there was a strong correlation, then the points would be arranged along a diagonal line – as the children's legs get longer so their jump would get higher. The tighter the points are clustered around this diagonal, the stronger the correlation; the more scattered the points are, the weaker the correlation.

Here we can see there is very little correlation between the size of the children's leg muscles and their high jumps. You can see that one child (Richard D) has the second highest jump (25 cm) but his legs are somewhere in the lower end of the muscle size range (29 cm). If you double click on a point on the scattergraph in *Textease Database*, it displays the record for that child – a useful feature.

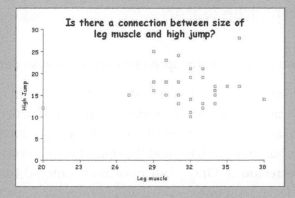

A *Textease Database* scattergraph showing high jump mapped against leg muscle circumference

By contrast, if we plot height against arm length, we see there is a much stronger correlation. The points are clustered much more evenly around the diagonal – the shortest person (James) has the shortest arms and the tallest person (Geoffrey) has the longest arms.

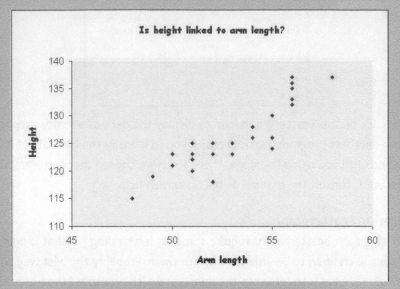

An *Excel* scattergraph showing height and arm length

How to form and test hypotheses

Hypotheses tend to be associated with science, but are just as important for statistics. Newspapers are inclined to report the outcomes of surveys and reports as 'facts' rather than hypotheses supported by evidence – some of which are more secure than others. A hypothesis is a 'guess' – but one which should be informed by evidence.

For this project, the hypotheses are principally based on correlations – e.g. 'the longer your legs, the higher you can jump'. In other words the measurements of leg length should co-relate to (i.e. have a relationship with) the measurements of jump. This sort of hypothesis is easy to test provided we have the relevant data. However, we should be cautious as we have only a small sample of evidence. Our conclusions ought to be qualified, e.g. 'For our class of ten- to eleven-year-olds, there is no link between leg length and the height people can jump'.

What will the children learn?

How and why to use a database (or spreadsheet) to analyse data

A key learning objective for this project is that children should be able to decide which form of analysis is the most appropriate to answer a particular question. By the time they reach the end of Key Stage 2 they will have produced bar charts and probably pie charts but may not have had the opportunity to compare two sets of data to look for correlations. Once data have been entered into a database or spreadsheet then the data can be examined and re-presented in a number of ways.

This flexibility and the speed and accuracy with which different forms of analysis can be carried out are the most significant strengths of data handling with ICT.

How to form and test hypotheses

Enquiry-led learning underpins the approach adopted in this project. Deciding which information needs to be gathered and then manipulating the data to find out if the initial question or hypothesis is valid is the basis for much of the National Curriculum and the National Numeracy Strategy. If you study the programmes of study for science, history and geography you will see direct references to an enquiry-led approach. It is less apparent in the other subjects but is, nonetheless, omnipresent – hence references to 'information texts', 'data handling', 'investigations' and 'problem-solving'. Constructivist approaches to learning assume that we continually form and test our hypotheses about the world around us. To achieve this successfully in maths, children need increasingly to develop knowledge and skills to improve the focus and accuracy of their suppositions and opportunities to confront and check their understandings. ICT, used carefully, can extend the range of what is possible.

How to interpret scattergraphs

As indicated in *What do I need to know?*, the children need to understand what the scattergraph represents to be able to make sense of the information it portrays. By plotting data about themselves, they will not only have an interest in the results of their investigations; they will be able to relate the points on the scattergraphs to particular individuals in the class.

How to conduct a statistical investigation

See Project 6 (*Statistical investigations 1*) for more information. This project starts by asking the children to make sense of someone else's data and then using this information as a comparator for their own findings. These approaches to data handling are mirrored in the real world.

Challenging the more able and supporting the less able: modifying the project for older and younger pupils

Modifying the level of mathematical challenge

The hypotheses that the children test can be modified to match their capabilities. You will need to balance the desire to let the children form their own hypotheses with the need to ensure that their enquiries are realistic and achievable. Children who are capable of more demanding work can be encouraged to devise more complex tests requiring more accurate data gathering. For example, they might investigate whether there is a relationship between eyesight and the ability to hit a target. Those needing more support might need to have their investigations structured more carefully and investigate more obvious correlations (e.g. leg length and long jump).

Adjusting the level of ICT demand

Give more responsibility for entering and sorting the data to those with more experience of ICT. Some groups could devise their own databases with subsets of data for investigating their own enquiries. As indicated in Activity 3, the responsibility for entering the data can be shifted to compensate for lack of experience – though this will reduce the potential for developing the children's ICT capabilities.

Adjusting the level of support

This can be achieved in at least three ways:

⊙ Using more adults such as teaching assistants or parent helpers.

⊙ Using peer mentors – the more confident working alongside the less experienced, though the mentors will need to be well briefed on their roles and responsibilities.

⊙ Adjusting the time and resources. Each activity listed above does not need to be taught in one session. Some activities could be taught over a prolonged period with, for example, groups of children working on an activity in turn on a classroom computer. Thus the new group could be guided by the previous group. Alternatively, if a teaching assistant is available, she could work alongside each group in turn.

Why teach this?

The organisation of the ICT programme of study statements for 'Finding things out' implies an enquiry-led approach – deciding what data need to be gathered, gathering (or finding) them, manipulating the data and interpreting them and deciding whether they provide the information required. Specific activities or resources are not specified, though some are suggested in the non-statutory guidance. Databases and spreadsheets are the most useful ICT tools for enabling the user to manipulate data which can then be turned into information. The process followed in this project is similar to that being used in the real world for the production of local and national statistics and in all branches of research. Attaching numbers to help understand phenomena is a fundamental principle of mathematics – which is reinforced here by making the numbers meaningful and interesting to the children who will be motivated to find the answers to their own enquiries.

The activities in QCA ICT Unit 5C: *Evaluating information, checking accuracy and questioning plausibility* provide few opportunities for the children to engage with the enquiries as they are decontextualised in an effort to make them applicable to many situations across the curriculum. Addressing the same learning objectives but in a purposeful mathematical context should help the children appreciate not only how but why a database is used as a tool to assist with the manipulation of data. This project can be used as a replacement for ICT Unit 5C.

ICT Unit 6D: *Using the internet to search large databases and to interpret information* focuses on the use of the internet to search for information. It requires the children to make judgements about the validity and reliability of the information they are finding. This project could be used alongside the activities in Unit 6D to enable the children to appreciate that issues of validity, reliability and plausibility apply to all forms of information and data – including their own. Use of the internet can be incorporated into a range of different activities and subject contexts and hence the learning objectives for Unit 6D are likely to be covered naturally in other areas of class work. Hence this project will focus more specifically on the use of the database in Year 6.

Data handling features prominently in the National Curriculum and the National Numeracy Strategy in which clear objectives are set for the gathering and analysis of data in purposeful contexts. Scattergraphs are not specifically mentioned in either document but both indicate the value of making use of ICT resources to '*interpret a wider range of graphs*'. While scattergraphs can be drawn by hand, the process is tedious when large amounts of data need to be entered. Using ICT shifts the emphasis on to interpretation and hypothesis testing by automating the process of graphing the results.

See also *Science* Project 6 (*Giant's hand*) and *Science* Project 9 (*Data logging*) for related activities.

Index

abstract representation/notation 14, 19, 25, 26, 38
addition 38, 110
algebra 120–1
angles 14, 49, 65, 84–93
associativity 119

bar chart 19, 27, 81
block graph 19, 27, 81

calculation 110, 112–22
classifying 25, 26, 35, 38
clip art 19, 21, 22, 106, 113, 118
commutativity 119
comparisons 18, 25, 26
conditional function/statement 109
constructivism 49, 131
control 10–16, 40–50, 84–93, 107
co-ordinates 86, 87
correlation 124, 129
counting 12, 18–27, 38

data handling cycle 75, 82
databases 38, 70–82, 124–33
directions 40–50
downloading software 44, 57

enquiry-led learning 131, 132
Escher, M.C. 54–5

factors 65
functions 112–22

gender 49, 70

hypothesising 71, 75–6, 124–33

iconic representation 35
image 21, 24, 79–80
information manipulation 26, 31
information storage 26, 31
installing software 33, 44, 46, 91
interactive whiteboard 2, 36
investigations 67, 107

language development 16, 25, 26, 38
line symmetry 54
LOGO 14, 41, 43, 47–9, 84–93

mazes 42–3, 11
measures 50, 67, 84–93
mental calculation 110
modelling 38, 66, 96–110
money 30
mouse control 31

National Curriculum 1, 3, 4–5
number line 11, 12, 16
numerals 16, 31, 50

one-to-one correspondence 38
online activities
    Alien Plants (Maths online) 108
    Balloon car (PBS kids) 100, 108
    Colin's coffee (NGfL) 99, 108
    Duck Builder (CGP Books) 100, 108
    Function Machine (Ambleside School) 112, 116
    Function Machine (Shodor) 112, 116
    I-board software 36
    Line Symmetry (Links Learning) 54, 57–8
    Loose Change (Oxford) 35, 36
    MSW LOGO 87
    Number Cruncher (Shodor) 113, 117
    Online pattern blocks 66
    Parachute game 99, 108
    Percy's Money Box (Neptune) 36
    Planet 10 (Planet Science) 108
    Plant Force (Cambridge University) 108
    Polygons Around a Point 52–3, 57, 66
    QuiltMaker 66
    RoboPacker (Houghton Mifflin) 52, 56–7, 66
    Rocket simulation (BigBus) 96, 98
    Roller Coaster (Funderstanding) 108
    Roller-coaster Designer (Cambridge University) 108
    Symmetry Game (Innovations Learning) 54, 58, 66
    Tess (Peda) 60

Tessellate! (Shodor) 55, 59
Tessellation Town (Math Cats) 55, 59
Time and Money (Learning and Teaching Scotland) 36
Totally Tessellated 53
Toy Shop (DfES) 37
West Point Bridge Designer 97, 101, 102, 107
operations 38, 97, 112–22

Papert, Seymour 48, 93
pattern 65, 66, 67, 109, 112–22
pictogram 18–19, 27
pie chart 82
plane (2D) shape 50, 52, 53, 55–6, 84–93
polygons 52–3, 66
Primary National Strategy 37
problem solving 38, 41, 48–9, 84–93, 122
programmable toys 10–16, 40, 84

QCA scheme of work 2, 6–7
quantification 82

reflection 52, 54, 65
relationships 82, 109, 112–22
role play 30, 32, 38
rotation 52, 54, 65

saving files 79, 86, 89
scattergram/graph 127, 128–30
searching 38, 78
sets 18, 19, 25
shape and space 48, 50, 67, 84–93
shopping 30–8
simulation 38, 66, 67, 96–110
software
    2count (2Simple) 20, 21, 23–5
    2Go (2Simple) 43
    aspexDraw 60
    BlackCat Logo 41, 43, 84
    BlackCat Turtle 2 43, 84
    Counting Pictures 3 (Black Cat) 18, 19, 20
    drawing tools (Word) 53, 59, 60–2,

Excel (Microsoft) 101, 103–7, 117–19
Granada Database (Granada Learning) 75, 76–8
Granada Draw 60
Granada Logo 43
Imagine LOGO (Logotron) 86–8
Information Magic (RM) 75
Information Workshop 2000 (BlackCat) 75
Junior Viewpoint (Logotron) 75
Let's Go Shopping (SPA) 30, 31, 32, 34
MicroWorlds (LCSI) 86
MSW LOGO 87
NumberBox 2 (BlackCat) 101, 103
Oak Draw (Dial) 60
Paint (Microsoft) 55, 62–4
Pictogram (Kudlian) 19, 20
Playskool Store (Hasbro) 30, 31, 32, 33, 34
Roamer World (Logotron) 43, 44–5
Starting Graph (RM) 19, 20
SuperLOGO (Logotron) 86–8, 89
Terry the Turtle (Kudlian) 43, 44–5, 84
Tessellation Exploration (Sherston) 60
Textease (Softease) 60, 75, 80
Textease Database (Softease) 75, 125, 127–9
Textease Draw CT (Softease) 60
Textease Spreadsheet (Softease) 101
Textease Turtle (Softease) 43, 44–5
Word (Microsoft) 53, 75, 80
sorting 18, 19, 25
spreadsheet 96–110, 112–22
symmetry 52–67

Tessellation 52–67
trial-and-improvement 85
triangles 84–5, 86
turtle graphics 40
2D (plane) shape 50, 52, 53, 55–6, 84–93

variable 96, 100, 101, 102, 107, 108, 109, 120–1
vector drawing programs/tools 59, 60–2, 65

writing frame 73

Printed and bound by CPI Group (UK) Ltd, Croydon, CR0 4YY

23/10/2024

01777692-0014